PRAISE FOR *THE VOICE OF MATH*

Dave has created a fascinating exploration of mathematics, its historical evolution, and its profound influence on our thinking and humanity. This book is filled with insights revealing the beauty, patterns, and infinite possibilities within the realm of numbers. A must-read for anyone intrigued by the hidden impact of math in our lives, whether they reside inside or outside of math professions.

—Aree Bly

Success Coach, Speaker, and Author of The Sustainable Brain

The Voice of Math is a book for everyone, regardless of where they are on their mathematical journey. Through Dave's insightful perspective on numbers, readers can explore their own understanding and appreciation for the beauty of mathematics. As you delve into the book, you may even discover another voice speaking to you, guiding you to your unique perspective. Everyone has a voice, and it is uniquely yours. I hope this book helps you find yours, so together, we can create harmony in the world of mathematics.

—Hannah Y. Park

Actuarial Manager I, Root Insurance

I find certain languages to be beautiful and various words bring me joy. Dave extends these concepts to numbers and mathematics in a very thought-provoking way. I have numbers that hold meaning for me and there are patterns in nature, especially in flowers, that I deeply appreciate. I enjoy numeric puzzles and often reach the end of my understanding. This book pulls all of these things together and helps me appreciate the history and beauty in numbers and mathematics. Novices of math and experts alike will find this a delightful read!

—Susan Osweiler

Chief Risk Officer, Sammons Financial Group

The Voice of Math is a great read for all but especially for those of us who love mathematics. As a mathematician whose research is mainly in Combinatorics and Algebra, this book accentuates the beauty of the integers and how they are such an integral part of our lives. Thank you, Dave, for the history lesson, as I can now use some of this knowledge when teaching my classes.

—Dr. Candice Marshall

Actuarial Science Program Director, Morgan State University

In *The Voice of Math* Dave Kester uniquely utilizes the concepts and history of mathematics to help the reader navigate through a thought-provoking journey of personal reflection, meaning, and purpose. Whether you are a mathematician, a scientist, or mostly disconnected from both, you will enjoy this book as an easily accessible and enjoyable way to look at our world and your life from a different, and hopefully more meaningful, perspective. Like an old and trusted friend, *The Voice of Math* will be a reliable resource that you will turn to again and again to help challenge your mind and give you some peace and clarity of purpose.

—Brian Rihner
Pastor, Grace Church of Denison, Iowa

First off, I am not what I would consider a math guy; however, Dave Kester has compelled me through many conversations (and now in his new book) to see the beauty of numbers and math. He has a unique way of using story to connect his audience to his lifetime pursuit of questions, observations, and creative solutions in the math world. When reading *The Voice of Math*, I am drawn to listen and to consider what I believe to be true about much of life and numbers. I can guarantee that interacting with Dave's book will challenge you in many areas of your life! And I appreciate that he has spent so much more time than I have researching and contemplating all things math ... it has enriched my life tremendously!

—Andy Fjellman
Owner, Jabbermouth Media

the voice of math

the Voice of math

UNLOCK YOUR
IMAGINATION
WITH THE
NARRATIVE
OF NUMBERS

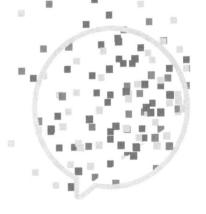

DaveKester

WITH MIKAELA ASHCROFT

Advantage | Books

Published by Advantage Books, Charleston, South Carolina.
An imprint of Advantage Media.

ADVANTAGE is a registered trademark, and the Advantage colophon is a trademark of Advantage Media Group, Inc.

Printed in the United States of America.

10 9 8 7 6 5 4 3 2 1

ISBN: 978-1-64225-954-4 (Paperback)
ISBN: 978-1-64225-953-7 (eBook)

Library of Congress Control Number: 2024910502

Cover design by Matthew Morse.
Layout design by Megan Elger.

Advantage Books is an imprint of Advantage Media Group. Advantage Media helps busy entrepreneurs, CEOs, and leaders write and publish a book to grow their business and become the authority in their field. Advantage authors comprise an exclusive community of industry professionals, idea-makers, and thought leaders. For more information go to **advantagemedia.com**.

*Dedicated to Julie, you are my trusted companion.
Your love and support are steadfast.*

*To my father, you taught me the invaluable lesson of
meeting challenges with courage and resolve.*

*And to my late mother, your passionate commitment to education
shines as a light of hope, especially in the dark spaces.*

CONTENTS

FOREWORD

By Charles Johnson, CEO, Actuarial Development Institute

We all seek happiness, but the journey can often leave us feeling lost and in need of guidance. Dave has been that guiding light in my life, unlocking the secrets to finding one's way over a lifetime. Through his business and his innate selfless love for others, he has taught countless individuals the path to fulfillment, authenticity, and happiness. The key to this path lies in learning math and logic. By understanding the interconnected foundation of the universe through the derivative of one foundational truth from another, we can begin to see the world in a new light. As this logical foothold is built, even the seemingly unintuitive and irrational parts of the universe become clearer. As grandiosity and immensity wash over you, your imagination is set free to explore. In this place of exploration, you can embrace your authentic self, love yourself, and find your place in the vastness of the universe. And, perhaps, in this journey of self-discovery, you may even find the transcendent.

In the quest for happiness, our journey is deeply intertwined with our interactions and relationships within the world we inhabit. This pursuit, a blend of seeking a place and forging connections, often leaves us navigating through uncertainty, in dire need of direction and wisdom. In my life, Dave has emerged as a beacon of guidance, illuminating the path toward self-discovery and fulfillment over the years. His endeavors, both through his professional ventures and his inherent altruistic nature, have enriched the lives of many, guiding them toward a life of authenticity, purpose, and joy.

Central to this journey is the embrace of mathematics and logic. These disciplines offer more than just a collection of numbers and propositions; they are the keys to understanding the fundamental truths of our existence. By delving into math and logic, we can decipher the code that underpins the universe, allowing us to view our world through a lens of clarity and insight. This comprehension builds a solid groundwork from which even the most perplexing and seemingly illogical aspects of the cosmos begin to make sense.

This exploration into the realms of knowledge is not just an intellectual exercise; it's a voyage of self-discovery. As we traverse through various fields of study, we uncover facets of our identity, gaining confidence and a deeper understanding of our essence. We learn to recognize our strengths and acknowledge our limitations. The awe-inspiring vastness that envelops us fuels our imagination, propelling us into realms of uncharted creativity. It's in these moments of exploration that we truly connect with our authentic selves, learning to love who we are and finding our unique place in the universe's expanse.

The journey doesn't stop at self-acceptance and exploration. It beckons us toward the transcendental, towards those eternal truths that remain constant across the ebb and flow of time and space. These immutable truths serve as a grounding force, reminding us of the grand scheme of things and our modest place within it. They teach us the beauty of humility and the grace of surrender, allowing us to release our burdens and find peace in the simplicity of being.

Through his guidance, Dave has not just shared a roadmap to personal happiness and fulfillment; he has opened the door to a profound understanding of the universe and our place within it. By championing the cause of logic and mathematics as tools for enlightenment, he has shown how, amidst the vast complexities of the cosmos, we can find simplicity, truth, and ultimately, a deeper sense of happiness. This journey of discovery, underpinned by the lessons learned from Dave, has the power to transform, offering a clearer vision of what it means to live a fulfilled and contented life.

We've seen amazing inventions in our generation, from the smartphone to GPS technology to AI. Imagine living in a day before these technologies where multiplication was required to navigate safely, and it was performed manually. For example, it was common to manually calculate something like $124 \times 3{,}998$. Repeating these types of calculations multiple times a day would not only be tedious but also include a risk of error. Then, in a spark of inspiration, you identified a clever way to transform this multiplication problem to one that can be solved by addition. Adding numbers of several digits is much easier than multiplying.

Take a moment and think about how we could ever make this leap. Why would you think we could even invent this shortcut? This was not given as a homework assignment to someone. Someone just had enough imagination to think this was possible. How would you even begin to solve this problem? This spark of genius is what changed transportation from a guessing game to a thing of certainty and opened the door for world travel. We still use this tool today in modern GPS coordinates, data compression, and dictating the f-stop on a camera lens. The tool is the logarithm.

In the latter half of the sixteenth century, John Napier had big plans to solve problems in astronomy and trigonometry. However, these tedious multiplication problems slowed his work to a crawl.

At Napier's birth in 1550, mathematics was still in its infancy. Calculus—the mathematical art used in all modern engineering and science—wouldn't officially exist for another century. The only way to work through complicated mathematical problems was to use the basics: algebra and arithmetic.

Other scientists of the time also felt the same. Scientists such as Galileo Galilei, Nicolaus Copernicus, and Leonardo da Vinci all used the same methods of solving equations, often resulting in their lives wasted in time-consuming calculations with a high risk of error.

True to the practice of his time, Napier kept in contact with other famous scientists and mathematicians of the period, such as Johannes Kepler and Tycho Brahe. Each astronomer had already made significant strides in calculating the natural motion of celestial bodies, but all agreed that, as new theorems and data were discovered, the calculations to prove them became increasingly arduous. Still, they all powered through, knowing the importance of discovery.

The desire to find a method of calculation that saved years of work became Napier's prime consideration as his frustration grew.

And, as all great mathematicians, John Napier imagined a deeper reality in the narrative of numbers.

The mysteries he saw in the universe could start with the simple need for a reduction in labor. So, the basic idea of logarithms was born.

Over the course of the next twenty years, Napier developed the first logarithm table. The table provided a shortcut to calculations as he successfully transformed complicated multiplications and divisions into simple addition and subtraction. The resulting log table spanned ninety pages with an additional fifty-seven pages that included extensive notes such as instructions for its use, a comprehensive overview of his findings, and its possible applications. At the time of its release, and even today, the creation of this logarithm table stands as one of the greatest mathematical achievements in history. This is only the tip of the iceberg of this story. I dive a bit deeper into this mystery in the Appendix.

There are thousands more stories like this about how numbers and math have arrived. Each individual story is interesting, but to me what is most interesting is the mega story of math. I don't view math as a subject. Rather, I view math as a growing collection of some of the best ideas we humans have devised and it is ordered in an amazing and systematic way. How did this tremendous collection weave together so seamlessly? Who oversees this process?

Math is a passion of mine, but it may not be a passion for you. However, you likely have your own passions. What drives that passion?

Let's consider rocks. When you view a rock, what do you see? We could view this rationally and talk about the weight, the size, the category, and other factual features that are true about rocks. This is how I think about rocks.

Fortunately, I have a friend John who rescues me from my narrow thinking. John is in the business of traveling the world looking for certain types of rocks. John knows all this information and more but when he talks about rocks he speaks with passion. I asked John what excites him about rocks. After all, they just sit there. Nothing could be more fixed and immovable as a rock, right? John is not content with just the outside appearance of rocks but is interested in what is inside the rock. So, he cuts the rocks and buffs until he finds these amazing colors that shine from within. He sees colors he has never seen before, all hiding within the rock. John sees this beauty as a personal gift for him to enjoy, which inspires a sense of wonder and thankfulness.

Figure I.1: *A rock I've received from John.*

I view math in a way similar to how John views rocks. There are a lot of features of math that are evident from the outside and things we can measure. This book is not about the outside information but the inside beauty. You may think math is as immovable as rocks. But math is a growing subject, which is evident in its history and context. As I witness the beautiful colors of math, whether it is the history or context or the math itself, I am, like my friend John, filled with wonder and thankfulness.

I want you, the reader, to experience the joy and beauty of math. This is a lifelong journey, but a good place to begin is the drama of how numbers arrived through people, culture, and time periods. My favorite thing is to tackle interesting problems and dive into the details. But that detailed math requires certain skill sets to appreciate. I want this book to be accessible for a wide range of backgrounds. I can't resist sharing a few math problems here and there, but I will try to make it accessible for those with limited math backgrounds with more detailed examples available in the Appendix.

Doing and enjoying math do require time and discipline. There are many forces fighting our pursuit of enjoying math beyond the obvious that it is difficult. There is always room for our human side to identify and appreciate true beauty and that includes beauty in math.

Another force that prevents us from enjoying math is viewing math only as a tool to apply. This is a one-dimensional perspective that leads to a race to the bottom. This pragmatic perspective soon converts the beautiful story of the logarithm function into only a button on a calculator. Then we measure success by how well we can use that button without understanding what it is. To allow more people to "succeed" with math, we pragmatically strip it of all its beauty, context, and depth and turn it into an empty shell.

One thing that makes math beautiful is its depth and width and many different perspectives. We're going to narrow our focus for this book to the development of the numbers. We will focus more on the history of how we developed our number system more than we will wrestle with the puzzles of the numbers.

Numbers may appear simple on the outside, but they are filled with mystery and intrigue with just a little slicing, dicing, and sanding. To appreciate this mystery requires an open mind and imagination. It may surprise you that math requires imagination, but I can't think of a subject that requires more imagination. You may view numbers as a closed book, but numbers are intriguing because they are not as predictable as we may think.

At its core, math challenges our notion of what we can know for certain and what we can't. This dance with certainty has made a profound impact on me, not only in math but also in life. For example, we're confident $2 + 2 = 4$ and equally confident that $4 + 10 \neq 2$. We know these facts based on simple counting and our calculator confirms it. But not all math is this simple and it is a humbling pursuit that doesn't reward closed minds. Math is an arena where we can constructively learn from faulty logic. What math teaches us is to step out of our perspective and into another perspective. It is like an exercise in empathy.

To me, faith is another easy step toward better understanding and appreciating math. Faith, like math, can be a thing of beauty. But if I pursue faith with a closed mind, I strip it of its mystery and miss out on the beauty. If there is not beauty in faith, what is it?

The connection between math and my faith doesn't end with an analogy. As I discover the height and depth of the beauty of math, I must ask myself, where did this beauty originate? This is the same question my friend John asks as he uncovers the beauty hidden in a rock.

This book is a story-based way to understand the people and context of how numbers have arrived to us. It is organized into two parts. The first part is the rational numbers, which includes all our integers and the fractions that we can generate from these integers. You can think of this as generating a ruler with the integer measurements and then the fractional parts within each integer. This is generally how we view numbers, especially in the context of the number line.

The second part challenges this notion of our concept of an ordered number line when we complete the set of numbers. To be honest, it's the second part where I get most excited because there is so much beauty, mystery, surprise, and wonder and there is a challenge to consider many perspectives.

Even though doing math is difficult, the further I travel on my math journey, the more I've realized that there is a kindness to math. For me, the kindness is like a quiet voice speaking to me to pursue more. The beauty of math is hidden, and it does not bang loudly. It does not demand that we seek it, but it rewards those who do. Sometimes identifying the patterns is only the beginning of the journey but understanding why the patterns exist is the real jewel. King Solomon shared a proverb thousands of years ago that inspires my journey to uncover this hidden treasure:

"It is the glory of God to conceal things, but the glory of kings is to search things out."

Before we begin, let's return to our previous math and ask, how sure are we that $4 + 10 \neq 2$? If it is currently four o'clock and we meet in ten hours, what time do we meet?

Let's open our imagination and let the journey begin!

P A R T   1

The Rational Mind

We like to pursue life rationally, especially in math. Poor logic always leads to incorrect results. Rational thinking can solve a lot of our math and life equations. Our calculator is a classic example of our rational mind. Quickly input integers, add, subtract, multiply, and divide. We can input decimal values. We even have buttons for e, π, and the square root button $\sqrt{\square}$. This is a rational way to solve math problems and seems to include everything we need for success. And, for our limited minds, it seems so.

CHAPTER 1

From Here to Eternity

God made the integers, man made the rest.
—LEOPOLD KRONECKER

Positive numbers seem like home on the number line. They've been with us since the beginning of civilization. In fact, you're probably so familiar with the concept of positive integers that infinity just seems like another step in the staircase upward. They are the basis for leading us from what we already know to most things that we don't. In essence, they lead us from here to eternity without skipping a beat.

Positive numbers offer the basis for understanding everything from simple mathematics to complex worlds. They're the basis for human growth. And they've grown with us.

But to those who don't delve into deep mathematical concepts, understanding how important they really are is difficult to understand. One, two, three … it's something we've "counted" on since before we can remember. In fact, you were likely influenced by positive numbers without even realizing it. Years ago, your mother or father probably held up their fingers to teach you the basics of fundamental mathematics. And you never knew it.

For those reentering the math realm, theory is likely off the table. You're interested in what you can see, not confusing graphs and unknown formulas. Despite what you may believe, I actually agree with you. If you want to become more interested in math, why would you start with something you can only see on a piece of paper? The best way to get started is to *visualize* it. So, before we go much further into abstract patterns, let's look at one you're probably familiar with.

Let's start by looking out the window. Depending on what time of year you're reading this, you might hear the *tap, tap, tap* of bugs outside your window. They're everywhere. And they've come from what seems like nowhere. In as little as a few days, your nice spring evening can turn into a mosquito-filled bonanza. It's a massive increase in bugs, something you didn't even notice happening around you.

Everything from mosquito population explosions to the mass extinction of the dinosaurs proves that the most common form of change is exponential. It's an interesting phenomenon to see such vast changes in short periods of time, but it's a common trend if you look for it. It's a pattern, and the first we'll touch on.

Of course, there are infinite other examples, but you might be more interested in something closer to home: the people around you.

If we were to set our zero marker at a theoretical position before written human history—approximately 10000 BCE, let's say—our pattern of human growth would seem fairly nonexistent. According to the resources of several researchers,[1] at 10000 BCE, the population hovered around 2 million. By 5000 BCE, roughly the beginning of

---

1   While we have multiple indicators from archeological finds to anthropological studies to mathematical models, some of the most noted archeological signs are here: Geoffroy de Saulieu et al., "Archaeological Evidence for Population Rise and Collapse between ~2500 and ~500 Cal. Yr BP in Western Central Africa," *Afrique, Archéologie & Arts*, no. 17 (November 22, 2021): 11–32, https://doi.org/10.4000/aaa.3029. Mathematical models to depict this are here: Douglas B. Bamforth and Brigid Grund, "Radiocarbon Calibration Curves, Summed Probability Distributions, and Early Paleoindian Population Trends in North America," *Journal of Archaeological Science* 39, no. 6 (June 2012): 1768–74, https://doi.org/10.1016/j.jas.2012.01.017.

written human history, that number jumped to 19 million, a staggeringly small jump in such a long time. The population reached 232 million at the beginning of the new calendar—0 CE—finally breaching the billion mark near 1800 CE. In the next two hundred years, the population exploded, landing at over 8 billion in 2023. And the number continues to rise. According to the data, the population increased more in two hundred years than it did in nearly 11,800.

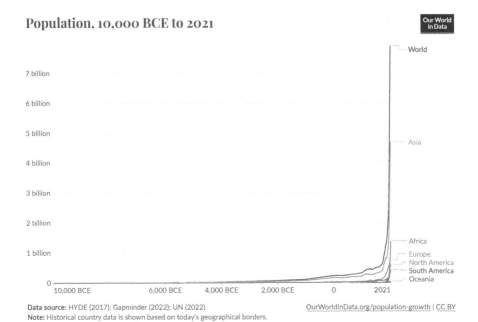

Population, 10,000 BCE to 2021

Data source: HYDE (2017); Gapminder (2022); UN (2022)
OurWorldInData.org/population-growth | CC BY
Note: Historical country data is shown based on today's geographical borders.

Figure 1.1: *Human population growth from 10,000 BCE to 2021 shows a pattern of exponential growth, despite taking millennia to take effect. Source: Gapminder - Population v7 (2022) and other sources – with major processing by Our World in Data.*

Of course, that's not really astonishing. Cleanliness increased vastly over time, and the introduction of new medicines significantly increased the likelihood of survival. Fewer mothers and children died in childbirth thanks to improved practices. In essence, the quality of life for humans has increased dramatically in the past two hundred years.

Surely this means that the slight nuances prove that there isn't a comprehensive way to understand positive growth, right?

Despite the setbacks and gradual growth, the human population still showed a pattern over time. In fact, when looking at integers as a mass collection, they all contain patterns that aren't always obvious to the naked eye.

So, what lies within positive integers that makes them so alluringly mysterious?

Ascending from Zero

Positive integers have a history as rich as art itself. It's the basis for the two types of mathematics: theory and form. Numbers take the form of much of what people consider mathematics today, and geometry is the manifestation of numbers. Every culture has had sense of one or both. It's the system of special awareness with higher functions—those associated with complicated math—that have truly defined positive numbers.

The thought experiments we take for granted today—such as the trolley problem or Schrödinger's cat—were hard-fought additions to mathematics. And, though we may not see them as such today, modern thought is a luxury made possible by positive numbers. For, without them, ideas wouldn't pay the bills.

THE LEGACY OF THE POSITIVE

The origin of numbers is tied directly with thought, meaning. Though some animals have a vague understanding of numbers, humans have tied them with conceptual understanding. In early history, numbers represented characteristics of the human body. A person's gender was intimately connected with a number: 1 was associated with male characteristics while the number 2 was linked to female attributes.

Cultures have developed language around numbers. Hebrews held such high regard of number symbolism, often referring to numbers in literature and speech. The number 1 represents the belief in a single God, 3 corresponds to revelation and strength, and 7 represents completeness, such as the completeness of the Earth after seven days.[2]

They're not alone in expressing religious symbolism. Many civilizations have found meaning in numerical representation. You might say that numbers were divine inspiration and counting came as a result of connecting humanity with something outside itself.

Cultural Number Growth

It's a silly concept, wondering where counting came into play. Humans have always used what they can see as the basis for numbers. Birds have two wings, dogs have four legs, humans have two arms and legs. And, as long as we've had fingers and toes, it stands to reason that we could count up to at least twenty. In fact, civilizations such as the Aztecs, Mayans, Basques, and Celts all used this method for base calculations.[3]

But more complicated calculations required tools. Archeologists have uncovered animal bones used to mark notches in early hunter kills, tallies likely used to differentiate kills between animals like

---

2   Other numbers also hold significance in the Hebrew language. Though translations definitely aren't perfect, part of the confusion many Bible readers face is the difficulty of understanding the significance of numbers. In some cases, the numbers recorded in the Bible don't refer to the numerical value at all but instead to their representation.

3   Early civilizations used "base" calculations to group numbers. The Babylonians used base 60, and we use base 10 to group numbers.

wolves and bears. You might say it was the first instance of accounting in primitive humanity. Yet, it's a method still used today. Tally marks on sticks, paper, or white boards all use the same concept.

The beginning for advanced mathematics, however, comes down to pebbles. Though its origins are up to debate, pebble mathematics became the standard for creating number systems. Indeed, the Latin word for *pebble* is *calculus*. Each shape of pebble represented a number grouping. For example, a linear pebble represented 1, a small pebble represented 10, a large ball represented 100, and so on. Other civilizations adapted this method using their own base numbers, transporting these stones where necessary. But there were problems with this method: the clay container holding the stones could break, and relying on human recollection is tricky. Since no rules explained the number system, a new method of writing, written on the clay boxes, depicted their contents. It became the basis for numerical writing.

With each successive generation, the need for arithmetic grew. Larger civilizations spelled the need for larger numbers in a system. Numbers that relied on counting, or even grouping numbers, was slow. Early civilizations struggled to maintain the pebble method as the primary way to manage counting.[4] Because geographically separated areas made it difficult to maintain a single number system, civilizations within Mesopotamia and Babylon integrated clay into the system. It was highly malleable and offered a way to regroup counting tools. It became a game of sorts.

Years later, the Egyptians were among the first to attach arithmetic to numbers. Each symbol, denoted in pictographs, held a place within a base 10 numerical system. By combining symbols together in

---

4   Early civilizations continued to use the pebble method within a clay pot to mark numbers, and anything that exceeded ten quickly became unmanageable. Writing excessive symbols on clay pots also became burdensome. Arithmetic was the easiest solution to the problem.

a long format, they created one of the first rudimentary examples of arithmetic. One of the first instances of Egyptian arithmetic recounted the riches a pharaoh took when conquering a city.

Separated by land and culture, the Egyptian base 10 system continued as the primary numerical base. In just a few hundred years, the power for addition expanded dramatically to subtraction, multiplication, and division, a perfect starting point for future influence.

In the East, Chinese mathematicians developed a similar system of numbering. Early representations of numbers were pictorial, and they followed the same base 10 method found in the Middle East. They also mirrored numerical representations, using addition and multiplication to denote large numbers.

Laziness, the true motivation for innovation, quickly changed the image of Chinese numerals. Long-form graphics became burdensome and difficult to read if written quickly, prompting the Chinese to develop symbols that significantly reduced error. The system's simplicity spread like wildfire, and the Chinese numbering system remains largely intact today,[5] a testament to its founders.

By 2200 BCE, hieroglyphics had entered Crete, the new center for advanced civilization. Though still using the same hieroglyphic structures as ancient Egyptians, the burdensome picture-drawing gave way to more manageable methods of counting, instead using dots, crosses, and bars to symbolize numbers.

---

5   The Chinese numbering system, created thousands of years ago, contained the same digital and decimal system that it does today. Though the surrounding countries—such as Vietnam—played a part in pronunciation and naming, the system became standard in the East. Even Japan, separated by ocean, developed a numbering system similar to rudimentary Chinese numbering.

Greek Revolution

The Greeks, the creators of geometrical mathematics, would rearrange ancient arithmetic altogether 1,500 years later.

Though math certainly served a purpose in ancient civilizations, its possibility for expansion left the ancient Greeks asking, why? This simple change in perspective altered the projection of math and history forever.

The patterns of basic geometrical shapes held similarities with measurement. And if we could manipulate geometrical patterns, it would only stand to reason that the same could be applied to mathematical concepts. Abstraction, the bending of reality to fit an idea, seemed only a natural consequence to mathematical design. For, if there were patterns in imagery, surely there were patterns in its mathematical counterpart.

In the early sixth century BCE, the concept of mathematical relations moved to basic structures. A triangle, among the simplest of shapes, revealed a simplistic pattern solved through an equation you've no doubt seen throughout your life: $a^2 + b^2 = c^2$. Attributed to Pythagoras,[6] the Pythagorean theorem served as a significant relationship between numbers, a pattern that showed consistency between rational numbers. This mathematical concept led to parallel hypotheses.

But surely there were more ways to manipulate math. Positive numbers were malleable, easy to shape. They were easily manipulated

---

6   I make the note that this concept is "attributed to Pythagoras." The truth is that the concept cannot be directly linked to him. Most scholars now suggest that he and his contemporary, Thales, are considered wise men of their time, but the concept now known as the Pythagorean theorem gained its name nearly five hundred years after his time. And yet, the theory was known hundreds, if not thousands of years before his time.

into geometric structures. What once relied on a straight numerical line could be changed to fit relationships in the physical world.

And so began Hippocrates's decent into ratios.

Among the most puzzling questions in early mathematics was this: How do you double the volume of a cube? At first glance, you may attempt to double all the sides. But, instead of doubling the volume, the volume would increase by *eight times*. Looking at the problem mathematically is the first step for many modern approaches, but Hippocrates only had the use of straight lines. So, the only way to create a solution is to shorten lines with respect to one another.[7]

Shape manipulation became the basis for understanding positive numbers. Numbers held a sort of power, a connection to each other that they'd never considered in the past. The relation between every number opened the door for explosive possibilities. If there was a connection between the angles of a rectangle, could there be a connection between the angles of a decagon? Or a pentacontagon? Or a centagon?

Each consecutive generation fought to add more knowledge, more substance, to mathematical concepts than the generation before it. However, the accumulation of much of Greek mathematicians is owed to Euclid's *Elements*. Within its thirteen books, Euclid explored different planes of geometry, formulaic expressions of geometric patterns, and the building blocks of rational numbers: prime numbers.

Prime numbers were among the first mathematical examples of abstract thinking. True manipulation of numbers surely meant that

---

7   The result is rather complicated, but here is Hippocrates's solution to the problem found in Archimedes's *On the Sphere and Cylinder*: "And it was sought among the geometers in what way one could double the given solid, keeping it in the same shape, and they called this sort of problem the duplication of the cube. And when they all puzzled for a long time, Hippocrates of Chios first conceived that if, for two given lines, two mean proportionals were found in continued proportion, the cube will be doubled. Whence he turned his puzzle into another no less puzzling" (Archimedes, Netz, and Eutocius 2004, 9).

every number, at its core, held the possibility of manipulation. The basics of geometry depicted the possibility of division within every segment. Visually, cutting the line into segments only made sense.

But certain numbers didn't stand up to the test. In a system composed entirely of whole numbers, it was impossible to cut numbers—such as 1, 3, 5, 7, and 9—in half without resulting in another whole number.

To our modern sensibilities, it's always possible to envision a number sliced in half; tacking on a decimal point is hardly a thought. And perhaps the existence of numbers that don't have a clear-cut halving point—in terms of whole numbers—is hardly noteworthy. But their concept resulted in a unique theory: these numbers represented the building blocks of all others. And soon a pattern emerged. Numbers had a foundational structure, and that structure persisted to infinity.

Among the most avid mathematicians, Euclid found a deep fascination in the descent into the unknown. His dive into number theory resulted in the emergence of a potentially limitless number of patterns found in positive integers alone. Geometry, it seemed, held a strong connection to all numbers on the positive whole integer scale. Odds and evens had their own patterns. In fact, the basis of mathematics had a pattern by nature.

So, if patterns existed at the heart of mathematics, surely any problem had *some* solution, right?

As the ancient Greeks dived further into math, their mindset changed. The nature of numbers necessitated a solution to questions that had a basis in reality. An accumulation of fairly simple facts relating to the nature of numbers meant theorizing. What we take for granted today, the scientific method, began as a revelation. A thought experiment nearly always had a logical solution.

And so, math separated from philosophy.

Philosophic thought didn't necessitate an answer, but mathematics implied that anything, conceived deductively, could have an answer, if proven. For nearly a millennium, the Greeks wrestled with mathematics, skyrocketing early scientific thought.

But the groundwork for modern mathematics came to a standstill between 270 and 275 CE. The Library of Alexandria, the storehouse for most of Western civilization's greatest written works, finally collapsed. The Library of Alexandria had endured amidst multiple fires and attacks over the course of several centuries, but its final decline was nearly absolute. The majority of written human innovation, including many ancient philosophical and mathematical texts, was lost to the ages and the fire that engulfed them.

From East to West

Struggles in the West prevented mathematical advancement. But, in the East, India's theoretical mathematicians rose.

India's early mathematicians formed the basis for the perhaps misnamed "Arabic numerals" we use today, a fact credited to Indians by early Arabic mathematicians. Alexander the Great's invasion of India spread Greek knowledge to the East, and records of advancements in astronomy and geometry slowly infiltrated history books. And, following their initial introduction, Indian mathematics flourished, developing concepts related to zero and decimal systems, algebra, trigonometry, and even calculus.

During the political upheaval occurring in Europe after the fall of the Roman Empire, the demand for mathematical advancement was low, and understanding of other cultures was significantly reduced. Wars fought within Europe remained within itself, only occasionally expanding to the Middle East in their fights with the Ottoman

Empire. An entire millennium faded away at the hands of internal wars. Interest in learning stood at the bottom of the list as survival became the primary concern.

However, mathematics continued to flourish within the walls of Arabic institutions. Taking notes from Indian mathematicians, Persians were among the first to develop a strong foundation in mathematics. Learning was a prized pastime, an activity of the most privileged.

The scattering of knowledge throughout the world slowed mathematics advancement to a languid creep. Disjointed bits of knowledge weren't enough to establish unified concepts. Social and economic issues became the most prominent problems to solve, and mathematics related to them was the only math that mattered.

Wars spread throughout the world, but knowledge came with them.

Though the war death toll only increased with the Crusades, the forced interaction between the two cultures encouraged the introduction of new ideas. Small pockets of knowledge survived, moving back to Europe seemingly slower than the march of the soldiers that carried them. But it was enough to spark a renewed interest in learning.

The final push of the 1400s to take the Byzantine Empire led to the loss of Constantinople, a previously Greek territory. In the wake of the Ottoman rule, Greek scholars fled to Italy, taking with them knowledge over a millennium old. Ironically, perhaps, the devastation of one intellectual haven led to the birth of another a mere thousand miles away. And with it, the birth of the European Renaissance.

Much of what we credit to modern mathematics came from scholars in the Renaissance. Though initially bogged down by politics, math flourished, creating an avenue for advanced thought. The symbols for addition, subtraction, division, multiplication, and many other mathematical signs were developed in rapid succession over the course of several centuries. Algebra adopted the alphabet, probability came into existence, and commercial arithmetic became mainstream, cementing positive integers into modern mathematics.

IN THE PLUS ZONE

Despite its colored past, positive numbers remain the foundation for mathematics. The patterns that emerge from positive numbers create the basis for all number theory. Odds and evens, prime numbers, and integers in general seem simple at first glance. It's why they provide the first proofs for modern science.

And there may be no pattern so simple as odds and evens.

Euclid explained that "an even number is one (which can be) divided in half" and "an odd number is one (which can)not (be) divided in half, or which differs from an even number by a unit" (Euclid, 2008). A simple proof, but it's the foundation for an analytical understanding of math. Children who can break down the logic of this pattern develop an understanding of correct and incorrect results.

True, the concept of odds and evens seems simply pedestrian to people who have moved beyond basic math, but what if we changed the concept to something a little more complicated?

Among the most common games associated with odds and evens is the game of chess. The classic game, likely existing since the sixth century,[8] utilizes strategic movement to beat opponents, but it all comes down to basic odds and evens. The board has sixty-four squares, and each player receives sixteen pieces. Some of the pieces receive more power the better the use of the board. Pawns, classed as the lowest, can typically only move one square (though moving two

---

8   For the history buffs, chess has only existed in its current form since the late fifteenth century. However, the game started as a form of military strategy, dividing each piece category into divisions of military, represented by the rook, bishop, pawn, and knight. It was first recorded in India and quickly moved to Persia, subsequently moving to Europe along with the knowledge of more advanced math.

squares at the beginning of their journeys is possible). The bishop can only move across squares of the same color, following an even number movement. Likewise, the knight can only move across four pieces, strictly even number movements. The most powerful pieces—the queen and the rook—can move without odd and even restrictions.

Figure 1.2: *Chess board.*

A simple play between a knight and a bishop can lead to hundreds of possibilities. The simplicity of the patterns that emerge within a chess game quickly elevate to complex strategies. After three movements, there are approximately 5,360 chess moves, and after four moves, there are nearly two hundred thousand chess positions.

The complex nature of patterns starts with simplicity and the overarching question: what if?

The mysteries of sequences, the emergence of patterns in nature, started the Italian Leonardo Pisano thinking at the turn of the thirteenth century. A rabbit farmer enlisted Pisano's help in solving an agricultural question: If the farmer has two rabbits, how many rabbits can he raise in one year? Given that the rabbits can produce a new pair every month and the new pair can produce another pair after their first month, the question gives rise to a possible sequence.

Because the above written pair in the first month bore, you will double it; there will be two pairs in one month. One of these, namely the first, bears in the second month, and thus there are in the second month 3 pairs; of these in one month two are pregnant, and in the third month 2 pairs of rabbits are born, and thus there are 5 pairs in the month; in this month 3 pairs are pregnant, and in the fourth month there are 8 pairs, of which 5 pairs bear another 5 pairs; these are added to the 8 pairs making 13 pairs in the fifth month; these 5 pairs that are born in this month do not mate in this month, but another 8 pairs are pregnant, and thus there are in the sixth month 21 pairs; [p284] to these are added the 13 pairs that are born in the seventh month; there will be 34 pairs in this month; to this are added the 21 pairs that are born in the eighth month; there will be 55 pairs in this month; to these are added the 34 pairs that are born in the ninth month; there will be 89 pairs in this month; to these are added again the 55 pairs that are born in the tenth month; there will be 144 pairs in this month; to these are added again the 89 pairs that are born in the eleventh month; there will be 233 pairs in this month (Pisano and Sigler, 2003).

The solution to the problem, though seemingly hefty in the translation from the book published in 1202, might be more familiar under another name. Granted, there are problems with the solution to the rabbit problem. For example, if a pair born to parents is two females or two males, they can't reproduce. The series—often seen in the form 0, 1, 1, 2, 3, 5, 8, 13, 21, 34, …—is the Fibonacci sequence.

You might be surprised to see physical representations of the Fibonacci sequence in other forms of nature. Flower petals, for the most part, come in 3, 5, 8, and 13. Pinecones and pineapples spiral in consecutive Fibonacci numbers. Seashells follow the same spiraling movement. At a much larger scale, galaxies also spiral in patterns consistent with the Fibonacci sequence.

Figure 1.3: *A seashell spirals in the same pattern as a galaxy, each following a form of the Fibonacci sequence. Though the scale for both is vastly different, they each follow the same pattern.*

It's a pattern we might not necessarily think about when first analyzing odds and evens. And it leads to an interesting idea. Is there such a thing as a perfect pattern?

As a matter of fact, there might be. And it all starts with odd and even numbers.

Do you recall, back in the days of your early education, the prime numbers? It's simple: they're numbers that can only be divided by 1 without leaving a remainder. It should come as no surprise that any number—except 2—that falls under the "even" umbrella isn't classified as a prime number. Prime numbers serve as an excellent basis for the creation of a number system. After all, every number can be broken into a set of prime numbers.

Prime numbers pave the way for more advanced mathematics. Perhaps to the modern sensibilities it seems as though chasing the next prime number is a fool's errand, or worse, a complete waste of time. Yet, they are the basis for modern cryptosecurity. For example, in a system known as the cyclic redundancy check—a method for detecting data corruption—computers use polynomial division over an infinite field of two elements to decrypt information, particularly that involved in online transaction security.

As a whole, we often consider "building blocks" to refer to the most basic of patterns, and that's true, but prime numbers offer another level of fascination. Surely there must be, at some point, a threshold where we would reach the ceiling of prime numbers, right? When numbers extend into the infinite, wouldn't there be some point at which all successive numbers would have more divisible returns than just 1 and itself?

Eternity is a foreign concept.

Though it's often hinted at in mathematical circles, it is, in fact, impossible to fathom. Everything we see has some sort of limit. A finality. So, proposing the existence of a pattern that exceeds beyond just the "very big" seems so foreign and, frankly, impossible.

The Positive Mystique

It should likely come as no surprise that, as far as humanity has come, it's clawed its way to this point. At first, the concept of time only extended to the life spans of those closest to the first people. As the concept of life outside of one's immediate point of view expanded over millennia, so did the realization that the universe we live in is much bigger than we'd originally imagined.

As the understanding of the universe grew, members of society vehemently opposed the idea of the vastness of the universe. In one legend, a prominent scientist delivered a message on the immensity of the universe. Cosmology, as was evident from centuries of study, extended beyond the solar system to stars in different parts of the galaxy, even outside of our own galaxy.

Frustrated at the insinuation, a critical audience member stood up and declared the information completely wrong. Instead of living in a world that was infinitesimally small compared to the larger galaxy, the reality was that the Earth stood on the backs of elephants who, in turn, stood on the back of a giant tortoise.

Puzzled, the scientist asked what the tortoise stood on, to which the audience member gleefully announced that it was yet another tortoise and another. Finally, the scientist asked what supported the tortoise on the bottom.

"Oh," said the audience member, "it's tortoises all the way down."

From the humblest of roots, positive numbers appear to be among the most simplistic of categories. Everyone from a young age learns to count.

But what happens at the infinite? Is it true that what we see is only the tip of the iceberg?

As it happens, the most basic understanding of numbers is far more expansive than we may have ever realized.

THE PRIME ENIGMA

Eratosthenes, a Greek mathematician living nearly 2,200 years ago, devised a technique to test the prime theory—the test of reaching an end point to prime numbers—using a method now known as the Eratosthenes sieve. For simplicity's sake, we'll use the following analogy:

Imagine a crusty prospector sifting for gold in a riverbed filled with ordinary pebbles. The sieve he holds only lets through the smallest pebbles, each of which can't break down further than they already are, just like prime numbers. The prospector takes the smallest of the gold pieces found—in this case, the smallest of the prime numbers, or the number 2—and sets it aside. As he inspects every additional pebble that is a multiple of the size of the initial gold piece, he subsequently throws them away. They are just larger stones.

He next moves to the smallest pebble that hasn't been thrown out yet, another gold piece. Again, he throws away every pebble whose size is a multiple of the gold chunk he's just found.

At the end of his search, he's accumulated a large pile of the most valuable pieces of gold.

Though this analogy's hardly a scientific delve into the nitty-gritty of the sieve of Eratosthenes, following the pattern laid out in the actual theory racks up a fair amount of gold. Removing all numbers divisible by the prime numbers is a fast way to find any corresponding prime numbers. But it's only realistic on a small scale.

And perhaps therein lay the fascination Euclid developed with this unexpected pattern. Prime numbers, at first glance, aren't easily determined. It takes lengthy calculations to discover new prime numbers. But, Euclid reasoned, if there is an endless sequence of numbers, there must be an endless sequence of primes.

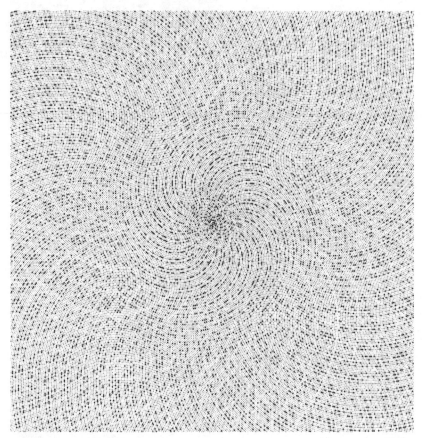

Figure 1.4: *The Ulam spiral shows a pattern that emerges by highlighting the prime numbers. Source: Bostock (2023).*

Analyzing prime number theories can become dry. The mathematical patterns seen as the numbers extend into infinity seem utterly basic. But do they graphically? As prime numbers extend into infinity, they develop a rare pattern that is depicted in an ordinary grid of numbers. If we place the number 1 in the center of a grid and then spiral the numbers outward in a counterclockwise direction, the prime numbers tend to form diagonal lines. The image is known as the Ulam spiral.

The prime numbers are unique when we consider the patterns within the numbers simply because they seem to not have a specific pattern at all, but clearly there is a pattern when we look at it graphically. At first glance, it appears that there are several main swirls within the graph, but when you look closer, you can see more patterns that emerge. And why is that? Well, we don't actually know for sure. There are ideas we have to give us clues as to why the patterns exist. If nothing else, it is an interesting perspective of prime numbers.

It's truly a mystery.

Throughout history, multiple mathematicians have struggled to prove patterns with the primes. At their most basic, prime numbers appear straightforward, but again scientists struggle with anything that isn't expressly proven. We're ultimately constantly driven by that insatiable question: *why?*

THE NEVER-ENDING POSITIVE NUMBERS

There is so much hidden in our positive spell. Simple counting makes the concept of positive numbers so accessible. It's something all cultures have utilized, to one extent or another. It's something easy to understand: putting one foot in front of the other. It's a method we've used for millennia.

Identifying patterns is the best way to answer one of life's most difficult puzzles: how far can we go?

Perhaps it's the hubris of man that makes us think that we can find a solution to nearly every problem. But what if the ingenuity we have, the drive to understand, is what makes us so unique? The patterns detected in history have inevitably led to astonishing discoveries and even greater understanding of what's around us. It provides a sense of comfort.

But the comfort of understanding drives an unquenchable need to ask *what next?* If we can take one step at a time to find the answers to life's most basic questions, how far can we actually go?

The Positive Side of Gravity

Perhaps the most puzzling phenomenon is gravity.

Everyone is familiar with gravity. It's often the cause of some of the most viral internet fails, a constant companion with all of us.

But aside from the casual reference to an apple and Newton, gravity provides a curious dive into understanding infinity. Gravity is the force that exists in every part of the universe. You can find the extent at which gravity pulls at any object using this equation:

$$F = G \frac{m_1 m_2}{r^2}.$$

If you're not interested in math, don't worry. We'll provide only the most basic understanding of this equation. The force of gravity is equal to the pull of gravity (a number so small that, if you were moving at a fast enough pace, you could jump from the Earth and float off into space) multiplied by the mass of two objects and divided by the distance between them. It sounds complicated, but let's think of it this way: a person jumping on the surface of the Earth will feel a lot more downward pull than someone who is jumping on the International Space Station.

The bigger the object, the more downward pull it will have. You can jump higher on the moon than on Earth, and—assuming you don't burn up on the surface—you couldn't jump on the surface of the sun at all … because you'd be squished into the surface.

Analyzing gravity at its most observable level seems straightforward. But it becomes far more twisted when the gravitational pull is so large that it bends the space around it. The force, which seemed so insignificant when observing it in space, becomes nearly endless when the mass is so large.

What if that small force was attached to a nearly impossibly massive object? Would it reach infinity? The most fascinating thought experiment lies in black holes.

They have so much mass that we don't know what's at their center. But, let's start at the edge.

At the circle surrounding the edge of the disk—the point known as the "point of no return"—is the last-ditch effort you have to leave the pull of the black hole. When you cross that line, a series of curious events occur. First, because gravity is greater at the center of the black hole and you enter face-first, your body will elongate. Second, time will slow down for you; because gravity can actually warp time, the people who are watching you descend into the black hole will think you're moving at an exceptionally slow rate when, to your perspective, you're moving at regular speed. And you'll continue to move toward the black hole forever.

From the perspective of the people watching you through a telescope, you'll appear to move toward the black hole indefinitely. From the time you pass the point of no return, you'll continue to fall forever.

For obvious reasons, no one has witnessed someone falling into a black hole. We don't truly know what's on the other side of forever. The limited experience we do have hardly extends that far.

Of course, there are theories. To some, a multiverse makes the most sense. The black hole could be a portal to universes outside our own. To yet others, it's a portal to go back in time. If time slows to nearly zero when approaching the black hole, who's to say that, on the other side of the black hole, the clock doesn't reverse?

Figure 1.5: *We don't actually know what's at the center of the black hole. While we can use math to predict what might happen past the point of no return, there's no way to know for sure what is beyond infinity.*

What we *don't* know is far greater than what we *do*. It's easy to get lost in the struggle to understand everything. And though speculating is a fun pastime, it's hardly a way to figure out some of life's more difficult problems.

Population from Earth to the Cosmos

But let's return to the Earth for a moment.

The explosive population growth we've seen in only the last few hundred years provides proof in patterns. Though there were several factors that limited a continuous growth pattern (disease and wars come to mind), the human population provides the perfect example of exponential growth. Like the graph at the beginning of this chapter, it's easy to see where continued growth would lead.

But what would happen if we took it a step further?

Much of what was once science fiction has now drifted into scientific theory. The 1902 French silent film titled *A Trip to the Moon*, the first science fiction film ever created, speculated what would happen if humans landed on the moon. The film suggested that a giant cannon could catapult explorers into space who were then greeted by heathen moon creatures.

It would take less than sixty years to reach that nearly inconceivable goal of landing on the moon. When astronauts reached the surface of the moon on July 20, 1969, they found mostly rocks.

Math is far more than simply numbers on a chalkboard or equations floating in the ether. It's the basis for understanding. It's the basis for curiosity, which is perhaps the greatest gift we could give ourselves.

It's math that makes up the difference we see in science fiction.

Math provides the steady arm at which we can create. And there's a beauty in that. If we hadn't observed how gravity affected the world around us nearly five hundred years ago, we wouldn't currently understand how we can improve vehicle performance, create a bridge foundation strong enough to carry many millions of tons, and even devise a machine to carry us out of this world.

It's the foundation of positive numbers—the numbers that started it all—that made it possible to understand what little we know about the universe. And it's the foundation for where we can grow in the future.

And it all starts with 1.

Positive Number Guideposts

Each set of countable numbers is full of patterns. There's a type of beauty, a type of joy, that comes from their simplicity. To me, the vast expanse only has one explanation: a Creator. Perhaps, then, it's easier to understand the positive numbers as more than the standard numbers we see in a number line. They're truly the guideposts that mark different perspectives.

Imagine running a marathon. Each mile holds a new view of the race. You won't feel the same way about reaching the first mile marker as you will reaching the twenty-fifth. It's a difference we all fail to see at times, maybe because it's easier to see how much you're suffering one mile from the end than the twenty-five miles you've already put behind you. It's only when you cross that finish line that you realize just how far you've come. Passing exams, taking that first step on your career path, and starting a relationship are all inevitable milestones in the human experience, and without them, I wouldn't be where I am today.

All these guideposts, including all the struggles associated with them, are beautiful. But they're messy, just like life. There are struggles, failures, and messes that are difficult to overcome. But with them are periods of joy. Without these struggles, we'd never know the joy. And sometimes looking back on these guideposts as the foundations for understanding who we are and where we're going is the only way to see them differently.

It's these guideposts that hold the foundations for viewing patterns. In positive numbers, there are intricate designs and simple shapes. In life, there are trials and consequences. The most obvious patterns include eating right, saving money, and doing taxes. But, experiencing joy or a lack of it requires a deeper look, a more subtle approach to patterns. Does the glimmer in a friend's eyes come from a selfless act? Will you feel the sting of regret if you don't take the chance to talk to that girl? Do you feel the hum of reverence when you stop to admire the sunset?

Those unflinchingly beautiful deep dives into your own soul often reveal thorns. The aches and pains from a loss of character development are difficult to approach. They're even more difficult to accept. But those parts of yourself you don't want to explore hold the key to your humanity. And, though they might not seem like it at the time, they are the foundations for the guideposts we eventually use to measure our lives.

As we'll see in the remaining chapters, positive integers are the foundation for the rest of the numbers. It's an intricate design that weaves the voice of math into the fabric of the most basic of human communication: numbers. This chapter gave you a taste of how these patterns can appear in nature, but it takes an ear to the ground to see the rest.

We're constantly pushing the boundaries of what we know. That push is a struggle, but the struggle is, perhaps, the most beautiful part of the mysteries of numbers. They often hold strange connections to who we are as people, though we might not know it.

Before you head off to the next chapter, just remember: Math can be messy. But it is a beautiful mess.

Conclusion

Our never-ending pursuit forward is the basis for not only math but life as well. But where did it all start?

Sometimes when we look into the night sky, we cannot see a single thing. It's blank. What does math have to say when there is nothing there? Why do we need a number to represent nothing? Can nothing possibly be interesting or beautiful? Does the voice of math have a sound of silence? It could be that no noise is the best noise of all.

Developing the rest of the mathematical numbers came with a struggle. Many had considered the positive numbers the only type of number necessary. All we see are positive numbers. And though it may seem like a logical beginning for us, one of the most difficult theories to accept was the existence of absolute nothingness. If we can't see it, it must not exist, right?

CHAPTER 2
In the Realm of Nothingness

And what are these Fluxions? The Velocities of evanescent
Increments? And what are these same evanescent
Increments? They are neither finite Quantities nor
Quantities infinitely small, nor yet nothing. May we not
call them the ghosts of departed quantities?
—GEORGE BERKELEY, 1734

Can something really come from nothing?

In the world of money multiplication, it often seems so. Placing a dollar in your savings account with the promise of a small percentage increase doesn't deliver much after its first compounded investment. But what happens when that dollar sits in that account, compounding yearly, for ten years? What about twenty? Or possibly thirty?

The current savings account interest rate is 0.53 percent. So, with a dollar sitting in your account for ten years, compounding annually, you'd only earn a measly $0.56. Still, you're earning something, even if it's a small amount.

Now suppose the amount invested were considerably larger. From a single deposit of $1,000, after a year with a 5 percent annual compounding interest rate, that $1,000 will become $1,050 in one year, $1,276 in five years, $1,629 in ten years, and $3,386 in twenty-five years. From a single investment, calculated over twenty-five years, the original investment plus the interest would reach approximately 3.39 times the initial investment. The increase in funds with the larger initial deposit and interest percentage is significantly larger, but there's still more at play to the example.

By increasing the frequency of the interest payouts to compound every three months, the $1,000 will become $1,051 in one year, $1,282 in five years, $1,643 in ten years, and $3,463 in twenty-five years. Increasing the frequency of compounding interest, the initial investment is multiplied by 3.46 times after twenty-five years.

So, the question begs: if there is such a significant increase in revenue because of a simple increase in frequency, what would happen if the frequency increased to monthly compounding, daily, per second? What if that number reduced to nearly zero?

The Evolution of Zero

Far from the gloried halls of the integers, zero has always been somewhat lowly. It wasn't introduced to the number line for millennia. In fact, it didn't seem necessary until greater mathematical minds began thinking. How do you define nothing?

UNRAVELING ZERO

Simple calculations didn't require a zero. Integers held the basis for counting, sorting, selling. Any astronomically large number wasn't necessary. In fact, for all human existence, until about 5000 BCE, scribes simply avoided the concept of nothing on the number line. It wasn't until the Mesopotamians put the Sumerians' concept of zero into written form, discovered as two wedges pressed into a lump of clay,[9] that the concept became more widespread.

Simple calculations were fine for a small civilization, but growing numbers made more advanced mathematics necessary. At its peak, the Sumerian empire boasted upward of eighty thousand people, so the Sumerians created more in-depth methods to calculate large numbers.

The complicated sexagesimal (sixty-based) system served as the basis for early advanced civilizations. Imagine the integers as simply the fractions of sixty, much as the numbers are on a clock. Though they didn't give a name to early depictions of the number ten, each division of six in the sixty-based system separated its final entry with a wedge. The wedges simulated the point at which the count would move to the next sixth.

After the destruction of the Sumerian empire, Babylonians simulated their work in mathematics by analyzing numeral notations found scattered across the desert. But, as time progressed, notations slipped. Some twenties mirrored the look of thirties, and some appeared the same as forties. Soon, it became difficult to separate distinctive numbers, but the pattern for zero remained the same. Every entry with a wedge figure represented some division of six. Basic methods for counting remained consistent with the original Sumerian design, but the changing "nothingness" theory drifted away for thousands of years.

Yet, the symbol persisted. Along with his plundering, Alexander the Great took another great prize from the Babylonians: the image of an "O." More than anything, it sufficed as a placeholder for the void, unfilled. Likewise, the Greeks who adopted this version of nothing-

---

9   Babylonians took much of what they learned about mathematics from the Sumerians, who recorded their numbers on clay tablets. For more information, reference *The Nothing That Is: A Natural History of Zero.*

THE VOICE OF MATH

ness represented it with completely relaxed fingers. The value simply meant nothing at all. The number had become a placeholder for large numbers, but philosophers certainly failed to understand its relevance to mathematics.

Hundreds of years later and thousands of miles away, the Greek influence on the concept of zero became an object for study within Indian mathematics. The Indians maintained the same symbol for zero that the Greeks had bestowed, but its significance changed. Instead of a mere placeholder for nothing, the Indians gave it a name. Suddenly, from an obscure concept, the idea of zero became much more than a simple depiction of the void. It became a positional notation.

Among the first to create the rules for zero, Brahmagupta used the notation as a starting point for astronomical projections. Calculating the size of the moon versus the sun couldn't start with one. In geometry, a starting point necessitated a value with no worth. In essence, true algebraic and geometric calculations couldn't survive without an originating point that contained no value.

Once Brahmagupta measured the heavens, algebraic requirements for the newly named zero became necessary. Adding and subtracting from both sides of an equation using known integers was easy. But what would happen should the quantity not exist? The question became more complicated when multiplying and dividing by zero. While the logic of assigning a value to a starting point seemed realistic, using arithmetic seemed divorced from pure geometry.

Across the pond, meditating on the same subject of originality, the Mayans developed their own definition of zero. However, instead of using the null value as the basis for geometry, the Mayans instead focused on its influence on time. According to their records, the beginning of time was dated August 13, 3114 BCE. Because the Mayans kept meticulous notes regarding time, counting down to their zero day was all but inevitable. So sold were they on the idea of starting from a null count that their monthly calendars began with zero. Careful deduction using their breakdown of solar patterns made the task easy.

Though it came easy to some, picturing zero was difficult. The Romans envisioned a number line with no "nothing" as the starting position. After all, when measuring three steps forward, there are four marks on the ground: the starting mark and the three subsequent steps. The music lovers have no doubt seen that in sheet music; a major third is two steps, moving from C to E, for example.

Increased suspicion of the theory of numbers outside positive integers created a roadblock for zero's progression into the Middle Ages. To quell the fear of adding zero to the number system, clerics were instructed to write out all numbers. The problems with addressing the concept of nothing bled into Roman numbering systems. And, though the concept of BC and AD in the calendar system we use today was devised in the sixth century AD, the year "0" still does not exist in the current calendar system.

Today, it's hard to believe a number generated so much distrust. But the greatest struggle those within the fight to add zero as a number to the system faced was an existential one. To ancient philosophers, after its naming, zero held the meaning of something and nothing at the same time. An impossibility, surely. The quandary became so intense that even the most avid mathematic visionaries failed to incorporate it as a bona fide number until the 1700s.

THE KEYSTONE OF SYMMETRY

However, by the end of the Middle Ages, mathematicians couldn't ignore the implications of zero. The various philosophies that led to the definition of nothing resulted in a single concept on which seemingly everyone could agree: zero was an origination. Zero is the starting point for positive integers. And, though they didn't know it at the time, it also represented the mathematical mirror that separated positive and negative integers.

Before lengthy mathematical proofs broke down coordinate planes and defined extensive relationships numbers had with each other, mathematicians broke everything down into the visual. And life, like math, has symmetry. The apple thrown into the air that falls back down, the marathon run that loops back to the starting line, and the pendulum swinging on a string all exhibit symmetry. The pattern of repetition is mathematically mapped.

The symmetrical nature of zero changed modern views of the physical world as well. Zero acted as the great equalizer, the point at which everything returned to normalcy. Fancy this: at what point does a blade of grass cease to move? The way it stands defines its equilibrium. When the wind blows, the grass moves either positively or negatively away from its standing position. However, it always returns to its original state.

Figure 2.1: *Image of a pendulum with its complete arc shown on either side. For the purposes of the example, assume it takes one second to reach either side from the center.*

The equalizing nature of zero, applied to mathematics, places it at the starting point for analyzing the delicate changes from equilibrium. At a point in which nothing is happening, brief changes, even those imperceptible to the eye, can elicit a large impact on nature. In essence, it is the butterfly's wings that may cause the hurricane half a world away.

To analyze these small changes, consider the pendulum. It swings in a perfect arc, supposing there is no outside friction to stop its motion. When mapping that on a graph, a single swing looks as it does on the graph. In mathematics, the motion is defined as $y = x^2$, as seen in Figure 2.1. Assuming zero is at the lowest part of the pendulum's swing, the motion looks completely symmetrical. We can capture its swing at any moment in time.

For the purposes of this example, let's assume the pendulum moves from the center to either side in one second. The bauble at the end of the string has no slope at the bottom, but it rises exponentially until it reaches its peak on either side. In a perfect world, the pendulum remains in constant motion, but that's hardly realistic. If left long enough, a true pendulum slows down over time, ultimately stopping. So, what if we calculate every incremental change in the time it takes to reach either side? At a nearly imperceptible level, the pendulum decreases in chunks we call dx.

Taking tiny steps in another direction, let's analyze something most people see with an Amazon logo printed on the side: a box. A box on the simple coordinate system centered at zero is symmetrical in all four quadrants. Let's take the corners to the extreme. That box can expand into infinity in incremental sections also called dx.

When most people think of the changes that can occur to a box, they think of the big picture. But changing the direction of the incremental changes may lead to something much more interesting than eternal expansion.

The opposite of zero is infinity, right? But what if it's not? What if, between the numbers of one and zero there lies another type of infinity? After all, how close can you get to zero without really touching it?

Approaching Zero's Infinite

The mystery of zero led many mathematicians to an early religious fervor. Zero depicted the struggle between good and evil. It remained the middle ground, almost an evil presence that split the world. Alchemy—a practice of the dark arts—includes the imagery of

Ouroboros laid out in a circle, the dragon that swallows its own tail.[10] From the fear it wrought, the fear of nothingness, was it a symbol for true symmetry or a mythical, abstract concept somewhere between the absolute and infinity?

While it hardly holds the superstition it did, zero does generate its own sense of wonder. After all, is it really possible to shrink something down so imperceptibly that it ceases to exist? Before the emergence of microscopes, believing that there was something as small as zero was nearly heresy. Today, do we give it a second thought?

Imagine again the box with sides with its center at the origin. When shrinking the box down by infinitesimal increments, the box becomes infinitely small, but it's not gone altogether. In fact, once it's out of the view of the naked eye, it enters the microscopic realm, then the nanoscopic, then smaller. The smaller the box becomes, the more difficult it is to measure it.

But, despite our inability to put a ruler to its sides, it still contains some dimensions. Shrinking the box down to as small as it will go using the latest technology will still never result in the box's complete disappearance. So, in the hidden realm between one and zero, there lies infinity, the infinitely small.

THE ZERO ENIGMA

Perhaps there was something to the speculations of zero's true nature. Of course, at the center of that large "O" there are no demons, no spells, but maybe there is a little bit of magic. Because, at the center of a zero, there is nothing. But there is no such thing as nothing. Even

---

10   The circular imagery found in alchemy and other "ungodly" practices and elements displayed an empty center. The idea of nothingness frightened many people, and it remains in many superstitions today. For more information, reference *The Nothing That Is: A Natural History of Zero.*

in the barest of space, there is *something*. So, in reality, it's impossible to completely conceive of something that simply has no space.

Identifying zero as a point of space at the origin was relatively easy: it represented a natural beginning. So, conceptually, adding, subtracting, and even multiplying by zero are simple. And, to an extent, dividing using zero follows the same rules. Zero multiplied by any number is zero. Zero divided by any number is zero. Multiplying and dividing zero by any number results in nothing because it's algebraically linked to the void. Zero colors any number with which it comes in contact.

But dividing by zero is a different animal altogether. Zero is still an ambiguous concept. In early interpretations of zero, dividing by it left the result unchanged. To discover simple mathematical rules applied to zero, mathematicians started with multiplication. In theory, the logic of viewing multiplication as a faster version of addition (five multiplied by five is adding five fives to each other) seemed a natural precursor to dividing by zero. If division was simply a faster version of subtraction, dividing by zero would take us into the negative numbers.

But that didn't quite add up.

If anything divided by zero was unchanged, $0/0 = 0$. But it doesn't. In fact, it's quite the opposite. The result of $0/0$ is undefined. One way to understand why it is not defined is that dividing any number into zero parts doesn't make sense.

But, what if the equation instead analyzes a different number? The concept of using numbers that aren't directly related with nothingness and their interaction with zero seemed a logical step in determining the new central location. Returning to the belief that division is simply fast subtraction, that would mean $a/0 = 0$. But again, that's untrue. Say, for example, we were to compare two equations using a positive integer as the numerator, $5/0 = 10/0$. Dividing by the zeros in the equation would result in the ridiculous conclusion that $5 = 10$, which is nothing but a bold-faced lie.

Though dividing by zero leads to madness, surely exponents wouldn't suffer the same fate. After all, exponents are just faster multiplication. They should be safe, right?

When interacting with typical integer numbers, exponents behave rationally: $5^3 = 5 \times 5 \times 5$. Using zero as the base for an exponent reacts the same way: $0^3 = 0 \times 0 \times 0$. But what does it mean when the positions are reversed? Does $3^0 = 3$? If 3 is multiplied by itself zero times, then it would equal itself. But, 3^1 also equals 3, which leads to $3^0 = 3^1$, or $0 = 1$, an impossibility. If we view exponents in terms of multiplication ($3^3 = 3 \times 3 \times 3$ and $3^2 = 3 \times 3$), it naturally means that $3^0 = 0$. If that's the case, $3^{-1} = -3$ instead of $^1/_3$.

One way to reconcile what a^0 means is to divide two exponentials, but we'll start with something a little more within the realm of reason, such as $^{3^7}/_{3^5}$. The resulting equation written out reads:

$$\frac{3^7}{3^5} = \frac{3 \times 3 \times 3 \times 3 \times 3 \times 3 \times 3}{3 \times 3 \times 3 \times 3 \times 3} = 3^2$$

If both the exponents were the same, the result would read $^{3^7}/_{3^7} = 1$. In other words, using division's rules, we know that any number divided by itself is 1. Applying that same logic, $^{3^5}/_{3^5} = 3^{5-5} = 3^0$, which is just 1.

Directly attacking zero exponents using division seems the only way to approach the problem of raising any number to zero. Because this method works for the 3 in this example, the natural conclusion is a rule that is the same for all exponents: $a^0 = 1$. If that's the case, then all integers, nay all numbers themselves, should follow that rule. The rule means $5^0 = 1$, $100^0 = 1$, and even $\infty^0 = 1$. The fractions follow suit with $^1/_2{}^0 = 1$ and $^5/_8{}^0 = 1$.

So, surely this roundabout way works for zero's own exponent. Following the previous logic, it's easy to assume that 0^0 would also equal 1, right?

Unfortunately, zero, yet again, presents a problem when raising it as an exponent. Using the previous method, the search for using zero as an exponent would look something like this: $0^5/0^5$, when divided, yields the following equation:

$$\frac{0^5}{0^5} = \frac{0 \times 0 \times 0 \times 0 \times 0}{0 \times 0 \times 0 \times 0 \times 0}$$

Once again, we run into a problem: dividing by zero results in the unknown.

It's obvious that there are elements of zero that are still difficult to understand, not only conceptually but mathematically. We never seem to know exactly how to handle something that simply doesn't exist.

It appears that any number associated with zero is given value. Considering the rules of division, zero reduces numbers to nothingness. In the case of exponents, zero gives all numbers a singular value. It would seem nothingness drastically affects mathematics, but to what extent? Does the result extend to infinity or cease to exist? Is it the nothingness that bleeds into other numbers, or is it something on the cusp of nothingness?

DANCING WITH THE EXTREME

In a world filled with *somethings* it's nearly impossible to imagine *nothings*. For scientists and mathematicians, though, it's impossible to leave it at that. If we can't define nothing, almost nothing must do the job, and Leibniz did just that by defining that old quantity we discussed earlier: *dx*.

Taunted by the idea that we could divide by zero, if only at the proper moment, mathematicians created the concept of "near zero," a number so incredibly small that we could skate by zero, almost touching it, to find a value that expressed near nothingness.

To understand how something might approach zero without hitting it was conceptually difficult. After all, how can you possibly know how to get that close to zero? If zero was a concept that took millennia to understand, "as close to zero as you can get without touching it" was far more confusing. But, as with the beginning of mathematics, understanding it by seeing a representation of it made the concept easier to swallow. It all started with a straight line. Finding the slope of the line determined its relationship to other points. So, by using the formula for a line, $y = mx + b$, we can determine what happens at any point along the way. This graphical representation made it easy to see what would happen if the distance between two points shortened to nearly nothing. A line, it seems, is a rather boring and uniform way to determine what would happen if we closed the gap to near zero.

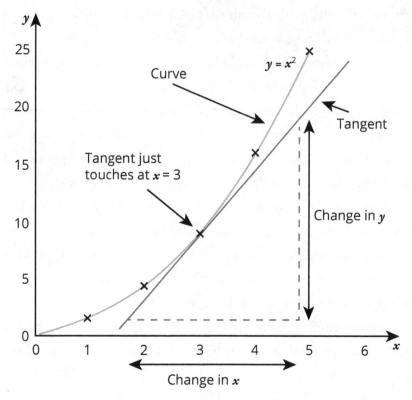

Figure 2.2: *The graph of a curve with the slope shown as the hypotenuse of a right triangle. The right triangle allows us to use the slope formula. Source: Resourceaholic (n.d.).*

So, ultimately, the journey led back to a graph that represented our example from earlier: the pendulum. It was the simplest symmetrical curve, and its slope is vastly different between two points along the curve. So, to determine the slope, the first solution is to calculate the rise and run of the slope at any point by creating small triangles with the hypotenuse at the curve, like that seen in Figure 2.2.

Analyzing a curve between two points only tells us so much, however. So, in the year 1666, Isaac Newton became fascinated with the idea of a single point. Of course, by a slope's very nature, it's impossible to find it without two points.

Figure 2.3: *a) Standing on a hill with both feet flat on the ground, you can feel the slope when you close your eyes. But, in b) if you stand on a pin, you can't feel any changes in the slope. The smaller the distance between the two points becomes, the closer we get to zero.*

Take, for example, a hill's slope. Let's say you stand with your feet flat on the ground. Your calves may feel tired after a while if there is a steep slope. If nothing else, you at least physically feel the effects of a slope. If you stand at a single spot, say, on your tippiest of toes, it's difficult to determine at how much of an angle you're actually standing. Closing your eyes and standing as straight as you possibly could, there's little chance you could determine if the slope is great or small.

Now let's assume that there is hardly any slope beneath your feet—as though you're standing on a flat bit of ground on the Earth's surface—and you can't feel the slope while standing on a needle that is making contact with the ground. Though it would take the skill of a master craftsman, you've happened upon a pin that would hold your weight without moving from side to side. You'd likely be unable to decipher just how much of a slope you're standing on. The size of the pin has become so imperceptibly thin that the distance between the two points is nearly zero ... but not quite. At what slope are you standing? You're on the Earth's surface, so there must be some slope, but it's barely perceptible.

If we can determine the slope of any point on the curve by calculating two points, what happens when the distance between those two points becomes infinitely small?

The question prompted one of the most famous mathematical disciplines in history. As we mentioned previously, Leibniz, a scientific contemporary of Isaac Newton, incorporated these small dxs into his version of calculus. For, he reasoned, the universe is made up of components smaller than the human eye can detect, though there was no possible way to determine the minute sizes at that time. Nature itself is broken into parts smaller than chunks infinitely small. So, no matter how far we break down the distance between two points in a slope, there is always some miniscule amount Leibniz coined as "infinitesimals."

Leibniz's view of these infinitesimals far exceeded a simple arbitrary concept. At its core, infinitesimals were numbers, smaller than anything ever known before. They were invisible, but vastly important to understanding what occurred at a single moment in time.

During Newton's dive into the mysterious dx, he insisted on a different name: "fluxions." Curves always moved in a fluid motion, so the smallest distances between them flowed with them, he reasoned. Unlike Leibniz, Newton's interpretation allowed for the fluidity of "next to nothing."

Though still undefined, it was the knowledge of these infinitesimals—or fluxions—that applied finesse to logic. For, whether dx corresponded to a finite number or was indeed fluid makes little difference. As Bishop Berkeley stated, these infinitesimals and fluxions "are neither finite quantities, nor quantities infinitely small, nor yet nothing. May we not call them the ghosts of departed quantities?"[11]

The point in their discovery was to test their limits.

Just as the point of a triangle or the border of a square has defined limits—the border at which the other line is directly met—any point in space should likewise have a limit, even zero. Because, if there is a point at which we almost touch zero without reaching its limits, we could understand the fabric of the universe. If there are theoretical *dx*s that reach a distance infinitely close to zero, could that apply to physical space?

Space Between Space

If Newton had chopped his famous apple in half, then in quarters, all the way down to the finest knife points, would he have reached the apple's basest parts? Unfortunately, the knife's blade still retains a certain thickness, even a slight one, making it impossible to break down completely. For, within the apple's makeup are molecular compounds, then molecules, then atoms, then even smaller.

But what is at the end of that line if we kept cutting? Would we ever reach zero? Looking at the problem theoretically, the question seems strange. How can you, even when breaking elements down as far as they can go, make something out of nothing?

The smallest particle modern science has detected is the electron. Nearly 1,000 times smaller than quarks—the particles that make up protons and neutrons—electrons constitute a foundational particle with a size smaller than 10^{-22} meters, though the actual size has

---

11   The criticism comes from George Berkeley who found the idea of infinitesimals an impossibility. It wasn't unusual for such fluid ideals to have serious criticism. They appeared to show a crack in early calculus. It wasn't until the twentieth century that logician Abraham Robinson provided a proof that depicted infinitesimals as logically consistent with other numbers, though it was a new type of number.

been estimated. Although the electron wins the prize as the smallest known particle, it isn't the smallest known quantity in the universe. Vibrating at the center of known physics is another quantity: the Planck constant.

Let's consider the pendulum again for a moment. The distance at which it starts from the bottom, then reaches the top of its swing, is the amplitude. Though the swings seem continuous to the naked eye, the tiniest time frame is actually divided into miniscule pieces. Of course, we've discussed this before in the form of distance, but this same notion applies to energy. Just as there cannot exist something from nothing, waves require some base energy unit, in this case, the miniscule number 6.626×10^{-34} joule-seconds.

It was this unique discovery, named after physicist Max Planck, that institutionalized the notion of wave-particle duality. In essence, all particles are both waves and have mass since only nothing can come from nothing. And in a breath as faint as a whisper, the idea of absolute nothingness was eradicated from modern physical understanding.

The Invisible Influence

When I look into the microscope of nearly nothing, what unlocks the imagination?

Point zero is often the biggest hurdle to overcome. As we've seen, nothing in the physical world is truly at zero, and neither are we. So, part of the challenge is seeing point zero. Something as simple as an inability to solve a math problem can seem like a zero moment.

The smallest moments in mathematics—from zero to the smallest derivation between zero and one—only have true significance because of the names given to them. Though the idea of nothingness has existed for millennia, the number "zero" didn't appear in mathematical equations until it had received a name. In essence, its name made *nothing something.*

During the creation of my business, naming it seemed the least of my troubles. Creating action plans, reaching out to the right people, even discussing mathematics seemed much more important than naming the business. The simple act seemed only a cake topper.

However, as I spoke with specialists in my field, most gave the same advice: consider naming your business as one of the most important parts of its creation. Little did I know that putting a name to it would open doors I hadn't imagined. The name changed my perception of the business. The name zeroed in on which tools I needed to provide to my clientele.

At first it seems like we need to ask permission to name something. Math teaches us we are free to create names and definitions at will. Then we live with the consequences. I still remember the time trying to name our online exam creator. Eventually, I chose Adapt and it is known industry wide. It still seems odd to talk to customers who use the word Adapt as if it is part of our world, like Google. But, that is the power of giving a name to something. It instantly feels more real. Of course, our parents had the same feeling when they chose our names. At first, it seems strange. But it becomes a part of our identity.

Unfortunately, names given in haste often stick. If they don't adapt, the names you give someone, something, or yourself can forever alter your perception for the worse.

In Louis Sachar's classic novel, *Holes*, Stanley Yelnats is sent to a summer camp in the Texas desert when wrongly accused of sneaker theft. There, Stanley meets Hector Zeroni, the quietest of the band of boys, who was given the name "Zero." Though the Zeroni name does share similarities to his nickname, it's due to Zero's perceived lack of intelligence that he's gifted the name. Though just a nickname, "Zero" dogs him throughout his time at Camp Green Lake.

Just as in the case of Hector Zeroni, labeling unjustly has often become the bane of many seeking to outgrow the names given to them during first impressions, youthful indiscretions, and false assumptions. The labels you're given by others may seem insignificant initially but can become the bat with which you beat yourself later in life. The simplicity of a derogative label may seem the only way to define yourself as you stop your own progress. The negative connotations associated with zero, developed over years of the fear of nothingness, seem only too realistic in life. The untapped potential, the unfulfilled dreams, the unrealized expectations seem only a weight when you seek to rise above your own perceived "zeros."

In times of despair, it's easy to focus only on the lack of progress. But, is that the true definition of your journey?

The names you associate with your success, the labels you create to define how far you've come in life, dictate your future success.

Looking from the outside in, analyzing infinity by "zeroing" in on nothingness is ludicrous. At the surface level, the steps it takes to reach infinity are astronomical. And starting at zero in our own lives is nothing short of painful at times.

But, how do we actually know that the starting point—our starting point—is truly at point zero?

The truth is, nothingness does not exist. Even in the deepest reaches of space, the universe contains some speck of mass or some fluctuation of energy. Even at the infinitely small, we are still far from true nothingness. For, if there were true absence of everything, any *something* would fail to exist.

When you start a new task, a new idea, a new vision, you're hardly starting from the basest point, for there is no such thing. You were given previous knowledge, helped by someone who has been where you are, or simply existed in a time when accomplishing your task was possible when it never was before. There is always some step above zero, some advantage that, if you narrow down your view of your situation, may seem almost imperceptible. However, when things seem their bleakest, their darkest, and the farthest possible from reaching your goal, you are still at least one step from oblivion.

The path to a greater purpose or the end of a daunting task can seem like from here to eternity, but every moment in time has steps, small though they may be, that propel you from one point to the next. Sometimes even the smallest movement—the dx—provides the only way to move onward.

Conclusion

We win the battle of zero by realizing we never really start at zero. We come into this world with an amazing body. We develop knowledge, skills, and abilities that we can use. We need only look back at the people before us to see more than our mere limited view of zero. Perhaps it's just family or friends who have impacted us. Sometimes it is reading about the history of math and gaining perspective and context of the journey that occurred before us. As amazing as Archimedes was, he did not have the gift of calculus we have. As good as Euler was, he did not have the technology we have. Yet, look what these people did. Going back in time seems like a natural thing just like the number zero seems like a natural name for nothing. Going back in time is a step into negative numbers from today, and the ideas and label of negative numbers certainly had their journey as well.

Unsurprisingly, the path forward often includes setbacks, dips, and even views into the past. Mathematics involves more than just a look into the positives. Though there's certainly more to the negatives than whole numbers, the question is, what is there to learn with a glance behind?

CHAPTER 3

Beneath the Surface

Negative numbers are utter nonsense.

—PASCAL, SEVENTEENTH CENTURY

Alone, in the abyss near zero, lies a cutoff point. Zero, for all intents and purposes, does not entirely exist, so shrinking a box down to its basest numbers as it approaches zero is the best we can hope for. The execution of *dx* is the closest we'll come to seeing the impossible in real life.

But how does the origin truly affect the rest of the world we observe?

Let's take a closer look into setting the origin. Life, as you knew it, began on the day of your birth. Your brain began its visual recording after you first opened your eyes. Though the world turned before you came into existence, that fact comes to you only due to others' experiences. In your own life, moment zero is so important it's still celebrated every year.

Everything that occurred before your birth plays a part in someone else's life. You may not have seen the construction of the Eiffel Tower nor the Battle of Waterloo, so how do you know it actually happened?

Well, what we think we observe may not, in fact, be the beginning at all. Our point of origin may seem like the obvious first step in understanding the world, but the world has gone on far longer than we've been in existence. For us, everything before our own origin may seem theoretical, but is it?

Since a date of birth is an arbitrary origin—especially considering that everyone that exists now and has always existed has their own—picking the origin is equally arbitrary. So, what's to stop us from choosing one at random? Last Tuesday is a good start.

Everything that happened from last Tuesday onward is considered in the positive realm. But what about what happened before last Tuesday? Well, it didn't simply cease to exist when we set our origin. So, what does happen when we look back beyond our moment of origin?

The Dark in the Abyss

Many who struggle with the concept of negative numbers today can find solace in the knowledge that, for millennia, brilliant scientists and engineers held such vehemence toward negative numbers that they remained in the realm of nonsense until roughly the last three centuries. As elementary students battle through the number line today, they'll scarcely know that philosophers fought valiantly to put them there.

Zero was historically elusive, even though the idea of "nothingness" has always existed, conceptual though it might have been. Negative numbers, though, hold a bit of majesty. They have always held significance in subtraction, but do they really exist at all? Or are they merely a mathematical tool?

For millennia, we've trusted only what we can see. As humans looked into the stars, they saw lights in the darkness. The way the sky moved, the way the Earth moved, the way the sun moved were all subject to speculation. All measurements made relied on positive integers. Negative numbers were a divergence from traditional thought.

Imagine, for a moment, conceding to the idea of nothingness. The small steps that are just shy of zero exist, obviously. But what if we took a step in the other direction? Could it be possible that there is truly something beyond what we see to the reverse?

THE NEGATIVE RENAISSANCE

As with nearly all numerical notation, ancient Babylonians paved the way for modern-day negative number notation, though they may not have viewed it as such. The sexagesimal system ignited the beginning of modern numerical recordkeeping, but at a terrible cost. Babylonian numerical records often became extremely cluttered. Numerical notations located too close to one another blurred into a haze. The resulting solution lay in representing numbers as fractions of sixty.

Negative numbers found a way into exponents.

Among the earliest records of Babylon, there exist negative powers such as $60^{-1} = \frac{1}{60}$, $60^{-2} = 1/3,600$, etc. Though the notation was expressed as a vertical wedge instead of a dash, they held the same powers as the negative exponents we use today. These fractions served to show infinitesimally small astronomical distances. However, after the destruction of Babylon, this method of utilizing negative notation faded.

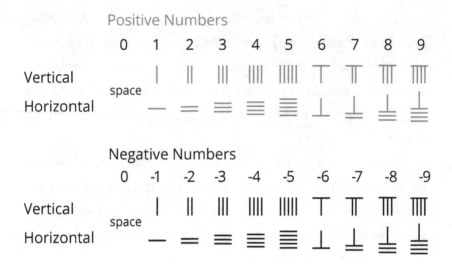

Figure 3.1: *Chinese figures were depicted as rods that implemented color coordination to do arithmetic faster. Positive numbers were displayed in red, while negative numbers were displayed in black. Negative number notation made it easier to differentiate changes in daily mathematics, which typically only referred to economics. Source: Sutori (n.d.).*

Negative numbers weren't much more than tools for simple arithmetic for the next few thousands of years. The negative numbers first appeared in Chinese mathematics as a method for calculating economics problems. Contrary to today, debt was depicted in black, while red depicted a positive balance. Though the exact date for the introduction of negative numbers into Chinese mathematics is fuzzy at best, scholars agree that it came into widespread practice by 200 BCE. Negative numbers were a natural conclusion to continued subtraction, and all were denoted using numbers similar to positive integers, as seen in Figure 3.1.

India's famous Brahmagupta, the author and theorist of the original zero theories, likewise used the concept of negative numbers in relation to debt. The introduction of zero gave Brahmagupta the means to add a place value at the correct location on the number system. Though preliminary, his definition of negative numbers was related in the following literary interpretation:

A debt minus zero is a debt.

A fortune minus zero is a fortune.

A debt subtracted from zero is a fortune.

A fortune subtracted from zero is a debt.

Even in its most rudimentary stages, negative numbers had a purpose. Their creation depended on a starting point, an origin. Though it took hundreds of years, new scholars found negative numbers useful in astronomy. Estimating a number above or below was denoted as "strong" or "weak," depending on its relation to the initial number.[12] And, though the emergence of negative numbers relied on a physical observation, the concept seemed relatively easy to grasp.

However, just as with zero, the world's most influential leaders in mathematics struggled with the concept. The Greeks, though authors of advanced addition, subtraction, multiplication, and division, struggled to reconcile the irrationality of something invisible to the human eye. After all, partitioning loaves—or even pieces—of bread, even in the most methodical manner, would never yield a negative number.

---

12 Estimating the location of a celestial body wasn't originally tied to negative numbers. Instead, slight variations in number approximations were seen in terms of positive numbers. For example, a planet may rest 5 units north of a star. Any estimation to either side of "5 units" was instead denoted by decimal points. Early astronomy didn't utilize the same methods of calculation as the Chinese used. So, instead of denoting variations in calculations, the term "strong" indicated that a measurement was higher than expected, while "weak" indicated that the measurement was lower than expected.

The ancient Greeks relied primarily on geometry to solve mathematical problems. And all geometrical problems consisted of positive numbers. Geometrical patterns served as the basis for arithmetic, and any "awkward" numbers were easily depicted as lines or areas within geometrical shapes. So, creating a geometric shape that depicted zero, let alone negative numbers, was impossible, right?

Diophantus, noted Greek mathematician and philosopher, took a dive into time on his quest to advance algebra during the second century CE. The knowledge acquired from ancient Babylonians denoted the existence of negative numbers used in conjunction with positive numbers. Surely, using this method, he could justify their use. By utilizing unknown variables in algebra, he could accurately create mathematical equations—such as the quadratic formula—without committing to introducing negative numbers into formulas. Genius, really, to create a scapegoat for introducing a concept widely accepted as impossible.

However, while recording his work in algebra in the book *Arithmetica*, he reached a problem: one of his equations, $4 = 4x + 20$, yielded a negative result: $x = -4$. Instead of analyzing the results, he simply called the result "absurd" and moved on.

Negative numbers' poor image changed slightly with the Persian mathematician Al-Khwarizmi. Instead of dismissing them outright, Al-Khwarizmi embraced the ideals of negative numbers, using them in equations in his book, *Al-jabr*, the foundation for the word "algebra." Algebra, he reasoned, was a science of impossibility and theory. Though a negative number couldn't exist within the realms of reality, algebraic expression provided a much more hospitable location for numbers that do not exist.

Persian philosophers grappled with the idea of negative numbers for centuries, perhaps inevitably circling back to mathematics from thousands of years ago. The Chinese and Indian representations of negative numbers in the form of debts slowly infiltrated the confines of difficult mathematical expressions.

A unification of cultural mathematics into a wholly accepted practice brought the theories of negative numbers to fifteenth-century Europe. Wars, conquests, and retrievals of fortune flooded Europe with treasures from Islam and Byzantium. However, like the zero, negative numbers were met with dismissive attitudes, if not outright moral opposition. While ancient texts translated into European languages simply ignored or held an apathetic view toward negative numbers, the fear of or disdain for using negative numbers that followed from the Greeks persisted.

More complicated mathematical equations—often those involving quadratic formulas—led Europeans to warm to the idea. Though many of the theories involved with negative numbers remained in the Orient until the nineteenth century, steady knowledge expansion gradually seeped into European mathematics. Only the need to account for debt and time into modern math theories convinced European mathematicians of the existence of negative numbers. They were an extension, at first, of the original number line. That explanation seemed good enough at the time.

In the mid-1300s, mathematician Nicole Oresme, likely influenced by the time in which he lived, correlated negative numbers to not only physical quantities, like temperature and velocity, but also to the soul. If negative numbers could define things with consistent change, why couldn't they also calculate the measure of pain and grace? Thoughts that were so reprehensible to the Greeks received new light when compared with religious texts.

But negative numbers went through their greatest renaissance when medieval thinkers utilized a method most scientists today consider second nature: creating problems to solve.

Galileo Galilei integrated negative numbers into physics. Following his house arrest for heresy, he delved into projectile motion, which created a convenient path for negative numbers to enter the number line: When calculating velocity, positive numbers represent the object in its upward motion, zero indicates its position at the peak of its trajectory, and as it falls back toward Earth, the velocity is then converted to negative numbers, as seen in Figure 3.2. It's a simple injection of symmetry into everyday mathematical equations.

Throughout the muddling Middle Ages, the attitude for negative numbers experienced ups and downs, all of which were eventually consolidated by scientists John Wallis and Isaac Newton. Finally, by the beginning of the 1700s, negative numbers were no longer being designated as "unuseful" and "absurd." Their integration into more advanced mathematics, such as calculus, necessitated their use.

SYMMETRICAL OPPOSITION

Following the wholesale suppression of negative numbers by Greek philosophers, many mathematicians rejected the idea of introducing negative numbers to geometry. And, in the plane of reality, their staunch refusal to acknowledge them has some merit. After all, have you ever seen an object that is truly less than zero? But because early Greeks were regarded as among the most intelligent and wise mathematicians and philosophers, negative numbers remained in a quagmire of disbelief for millennia.

Until René Descartes's Cartesian coordinate system[13] gained traction.

The new system took a unique approach to negative numbers, visualizing a viewpoint around a central location. There was an equal back to forward. It combined analytic geometry with theoretical algebra for the first time, but its creation took a mathematical renaissance, emerging for the first time in Descartes's 1637 work, *Discourse on the Method of Reasoning Well and Seeking Truth in the Sciences.*

The visual representation of negative numbers suddenly opened the door to negative numbers in mathematical equations. The "utter nonsense" negative numbers represented in theory became plain when represented on a number line. All it took was a change of perspective.

For millennia, mathematicians had used subtraction, like $8 - 5 = 3$.

So, what happens if, instead of seeing 5 as a positive number, we assign it a negative value and add the two numbers together? The equation would instead read $8 + (-5) = 3$, and the number line would instead look like the following.

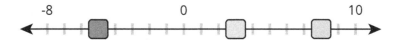

---

13   Descartes based his Cartesian system on philosophy and skepticism, all of which was largely based on religion and new information gained within the burgeoning Scientific Revolution. Following Copernicus and Galileo's work studying both mind and matter, Descartes adopted the ideas of duality in everything from philosophy to mechanistic laws. Though it may seem simple today to believe that because there is an up there is also a down, the real struggle lay in the definition of an origin, which, as we've learned before, many struggled to accept.

This simple change of thought inspired the acceptance of negative numbers. Seeing it simply as a shift from the center made negative numbers easier to grasp in addition and subtraction.

When we use negative numbers in multiplication and division, it's like they change the rules of the game. Normally, when we multiply something, it's like we're adding the same number over and over. For example, if we have 5 times 3, it's like saying 5 + 5 + 5. But, if one of those numbers is negative, like in 5 times –3, it's as if we're doing the opposite of adding. Instead of getting more, we're taking away, so it's like saying –5 (take away 5), then again –5 (take away another 5), and once more –5 (take away yet another 5).

Now, with division, it usually feels like we're subtracting something repeatedly. So, if we have –15 divided by 5, normally, you'd think of dividing as breaking something down into smaller parts. But here, because we're dealing with a negative, it's like we're doing the opposite. Instead of taking pieces away until there's nothing left, we're adding in a backward sort of way. It's like we start at –15 and then go backward by adding 5, three times (–15, then back to –10, then –5, and finally to 0). So, instead of division making things smaller, with negatives, it's like we're moving backward to get to zero.

Though these rules seem convoluted, they all center on symmetry around the number line. If there is a number 3, is there not something of equal magnitude in the opposite direction? And though difficult to grasp, negative numbers tell us twice the story positive numbers do.

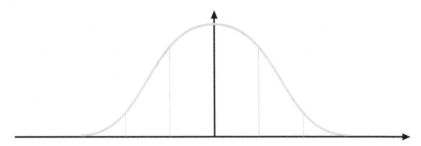

Figure 3.2: *The bell curve is a near-perfect representation of the flipped formula of $y = x^2$. The center line indicated here shows the median values, but what if it instead represented our center of origin?*

Take, for example, a normal curve.[14] We've all seen one: they look like a bell, and they're commonly used in statistics to represent what a population looks like, often looking like the one in Figure 3.2. When displaying the bell curve, nothing displayed is negative. Even though there is symmetry in the bell curve, we're not relying solely on the left-hand side of the curve, or the negative numbers, to see the whole picture.

But what if we moved the origin?

Let's reexamine the pendulum. The pendulum's swing hit zero at its lowest point. Anything to the right of its lowest point is positive. For the purposes of basic analysis, that may be enough. But it's only half the story. The negative arc of the pendulum didn't cease to exist simply because it resided on the wrong side of the swing.

---

14   For those of you struggling to remember what they were for in math class, think of the classic example: test grades. The test grades for a class range from 0 to 100, and assuming that all the students in the class have studied differently, the results would vary wildly. However, statistically, most of the students would land in the middle of the grading scale (represented in the peak of the curve), with some students excelling and others failing. The sides of the bell curve represent the outliers, and getting a test grade in those ranges is improbable but not impossible.

Far from the "unuseful" moniker given to the unfashionable side of a pendulum's swing, negative numbers are, in fact, necessary to see the whole story. Without them, our perception of not only a minute ball on a string but a whole universe may cease to exist. For nature requires symmetry, and there is none without negative numbers.

Beyond the Horizon of the Negative Infinite

This revelation of symmetry is not only beautiful but mathematically necessary. Unknown variables on one side of an equation aren't solvable if there is nothing on the right side of the equal sign. In nature, we couldn't know the true nature of a toad if all we saw was half of it. And, far more theoretically, we could have never understood the creation of our world had we not known and recorded a visible process, limited though it may be.

So, what *does* that "wrong side of the swing" of a pendulum tell us?

In the case of a simple pendulum, the symmetry of motion defines how high the pendulum will swing. Using the small increments of dx, we can even determine if there are changes in the swing. Tension in the string, even the slight rubbing the pendulum line experiences at its peak, can gradually degrade the movement. Each important feature tells us the true nature of so many things, like how gravity affects movement, how quickly friction can stop its motion, and how close you can put your face to the pendulum before it hits you.

It all relies on that negative swing, that opposing side that, though theory may have taken thousands of years to establish it, has been in the darkness behind us all this time.

THE OPPOSING RULE

Of course, there have always been forces we can't see, things beyond our understanding until they were written down or examined by someone else. There is always an equal and opposite force, a balance of equations, an invisible energy we may not see but that we certainly feel. For, with the passage of time, the elements of the universe that have collided to provide what we can and cannot see have all had a history we cannot always determine.

We know from scientific discovery of the existence of an equal and opposite force to every action. While standing on the ground, the pressure you exert on the Earth is responded to in kind. It's why you don't fall through the ground. At the galactic level, centripetal motion forces all the stars to extend outward while a strong gravitational force keeps them locked into a single galaxy.

A problem develops when we delve deeper into the abstract. Where, then, do we draw the line? If the center is arbitrary, if it moves depending on either an invisible line or a theoretical location, how can we truly know what enters the negative realm?

There are only two conclusions we can draw from this claim: negative numbers are either invisible or impossible. If they are impossible, then there is nothing truly beyond zero. There is no history, for there is nothing beyond a single starting point. But if negative numbers are invisible, they are nearly magical, for they are truly our creation.

But identifying what lies beyond that line is sticky. We can use symmetry to determine much of what is only barely in the negative realm, but that method doesn't always work. Sometimes what we think we see is actually an illusion.

A common example is the basic equations: $y = x^2$ and $y = x^3$. Placing our demarcation line at the center of each graph results in a picture that looks relatively the same. Both increase exponentially over a short period of time as we view the graphs with positive x-values. Both follow the same relative path. But the similarities end when we consider the negative x-values.

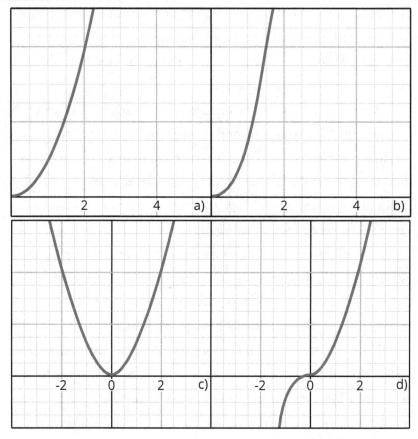

Figure 3.3: *These graphs show how we can often rely on symmetry to determine what is beyond the zero mark. a) The equation $y = x^2$ is shown from zero into the positive numbers. b) $y = x^3$. Both look similar (though not identical) to each other when viewing only the positive quantities. However, when we zoom out to see the negative values of equation $y = x^2$ (depicted in c) and $y = x^3$ (depicted in d), we see the inverse of what is seen in the positive plane.*

While the equation shows $y = x^3$ does have its own form of symmetry, we can't always rely on what we initially see. There are always clues to determine what is in the negative realm, but reliance on data becomes the most reliable way to determine what is beyond the veil.

THE INFINITE DEPTHS

What lies behind what we can't see—or, perhaps more importantly, what we can't extrapolate from our current view—stretches to infinity. It must be so. If there is an infinity linked to positivity, negativity must follow the same pattern.

But things get fuzzy the further we go down. After all, negative numbers have always been subject to mystery.

Imagine standing at the edge of the ocean. Modern nomenclature has already determined that sea level is at zero. Below you, curling at your feet, is water that is roughly at sea level. You can see everything below your feet at this point. However, as you progressively move further out to sea, you begin to lose focus on your feet. The further you go, the less you can see from above. By the time you're up to your neck, you can't see your feet in the swirling water below.

At this point, you begin to swim. As you make it a mile out to sea, all you can see below you is blackness. To your field of vision, there is no end to what is below you. For all you know, there is no bottom. There is only more ocean spreading out before you.

Though we may theoretically know there is a bottom, an ocean floor, the water becomes more mysterious the further we get away from the shore. Though we can see a taste of what lies beyond, it's never fully in our view. The further we are from what we know, the less certain we are of what lies beyond.

Figure 3.4: *The further we go below zero, the less certain we become of what lies beneath, just as it may seem as we drift along in the ocean.*

The first step is that small *dx* away from zero.

Since something smaller than absolute nothingness requires brain power far greater than we need at the moment, let's create our own center, or our own position zero. But, because stretching the mind is a favorite pastime of ours, let's assume that the center line, or the zero in this case, is the beginning of written human history, some five thousand years ago. We can't truly know what occurred at 3001 BCE, but we can extrapolate based on early writings and evidence found in the landscape—tossed tools in pools of water and made out of a variety of materials that, using carbon dating, reveal their age and relative time frame in which they were deposited. Just one year outside of written history is relatively easy to extrapolate. Yes, it's theoretical, but patterns of human behavior make it easy to extrapolate what we can't see.

But what if we extended that point further?

Dates and definitive history become but a blur when the creation of a large asteroid, the likes of which destroyed most life on Earth, occurred sixty-six million years ago. Though no history was left behind to analyze the impact of the asteroid, there were clues. The Chicxulub crater left an indent seven to nine miles wide in the Yucatán peninsula, leaving behind iridium—a mineral typically only found in extraterrestrial bodies—and crushed clay layers. The impact of the body likely raised a dust cloud so gargantuan that it nearly blocked out the sun. Plant life failed to grow following the devastating cloud, disrupting Earth's climate.

The raging debate created from theories of life's decline on Earth around sixty-six million years ago is a great indicator of the instability of our knowledge. Carbon dating of bones and the iridium left behind give us clues, but only theory based on observed science can draw us to the conclusion that this asteroid killed the dinosaurs.

The complete devastation of a species, long before humans ever roamed the Earth, is one of the greatest scientific discoveries ever made. But it's not quite a discovery. Symmetries with modern events—like the destruction created by humans' most powerful weapon, the atomic bomb—can give us a glimpse into the past. The mathematical and scientific analysis of evidence left behind many millions of years ago can provide answers.

But there's more. Reaching back further from the most recent complete devastation of the planet, we have the planet's creation. The fact of the creation of the Earth is obvious, but the *how* and *when* are more hotly debated questions. Far less is understood about this time, but based on astronomical observations, we've come up with some theories.

According to the theories, nearly 4.6 billion years ago, a cloud of dust and gas floated about at the edge of our galaxy. Over the course of millions of years, these clouds gravitated toward one another, eventually forming the sun. The upper atmosphere of the new star repelled all heavier gases, pushing them into orbit. Each heavier element continued to crash into the others, forming bigger, more dense materials. Rock formed. Planetary spheres began to take shape. Earth's magnetic core solidified before its large gravitational pull forced more rock upon its surface.

During its young tenure in existence, a large impact collided with the surface of the new planet, catapulting the current moon into space to forever orbit the Earth.

As friction on the surface of the planet became more raucous, volcanos formed to alleviate the pressures that lurked beneath the surface. Smaller, more consistent collisions with the Earth resulted in water deposits. And, as the Earth took form to eventually support life, the crushing violence of the past stilled. The atmosphere would primarily guard against most future problems created as Earth continued its long development to our time.

But, again, this is just theory.

We can replicate the creation of matter on a minute scale by using particle accelerators to fuse elements together. We can see the violent collisions in the sun to determine how fusion takes place. We understand that gravity, a weak force that pulls objects together, can create planets. We know there is water on the surface of comets.

The age-old question of the Earth's birth year has pervaded for millennia. And though its estimated pop into existence is a mind-boggling 4.6 billion years ago, that number continues to move backward with each emerging theory. And though there are conjectures of the Earth's creation from small replications to limited observations, the true creation of a planet, let alone one that can sustain life, is likely forever a mystery.

Using milestones like the beginning of written human history or the creation of the Earth or even the universe is a large jump. Assuming we have enough small steps from which to derive our conclusions, we can predict the invisible.

And though the information derived from the creation of the universe is truly fascinating, the process also works in reverse. Using symmetry, we can predict the future from the past.

Again, the ever-persistent equation $y = e^x$ rears its head. If you'll remember, e^x represents the exponential growth many processes experience. But, curiously, it doesn't end at zero. Exponential growth continues into the negative realm. And, just as we could see the vast changes as we approached zero from the right, we can use the same processes when viewing the equation from the left.

So, what lies within the first minute steps backward, between –1 and 0?

Well, infinite growth past the zero line in the positive goes upward at an exponential rate for eternity. However, when we move the opposite direction, exponential growth decreases rapidly.

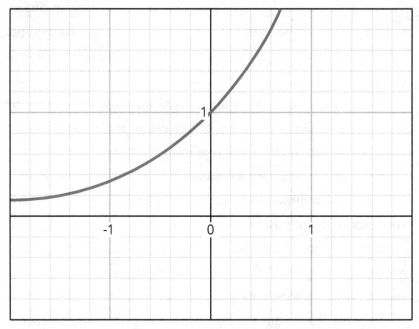

Figure 3.5: *The equation $y = e^x$ shows exponential growth. But, when time is zero (which we can define as the current time), the equation equals 1. So, the negative elements of this equation provide the perfect basis for us to utilize the past to understand the future.*

The curve flattens out quickly over this span, but it remains in decay. The equation shrinks closer to zero on the *x*-axis, but not quite.

As we expand those limits further, surely the equation would equal 0, but at what point? Moving back to –10, –20, –30? Each of these values brings us closer to zero without actually touching it.

The quest for a conclusion to infinite decay doesn't exist. It doesn't matter how far back we look on the graph. At one point the thickness of the exponential line may appear to line up with the *x*-axis, but it never really does. It moves backward for eternity, kissing the line but never coming to a satisfactory end.

Which begs the question, does negative infinity really exist? Is there some magical point at which we can go back on the *x*-axis far enough to actually reach an end?

In mathematics and science, exceptionally large numbers in the positive or negative direction are simply labeled "infinity," though they technically do not qualify. To truly hold infinite status, numbers must extend without end. And, the truth is, perhaps infinity is too difficult a point to grasp.

If we travel back in time 13.8 billion years ago—truly a number that should qualify for negative infinity status—we come to a finite point: the beginning of the universe. But is that truly the beginning? From what we know of the universe, it is expanding at an accelerated rate, but what's to say that the expansion of a previous universe didn't create this one? Perhaps the motion of our universe is just a single part in a truly infinite plan?

In the Realm of Minus

Perhaps it's the pattern that keeps us coming back to the numbers. There's a truth in patterns that isn't swayed by personal opinion or circumstances. Truth is truth, regardless of where it comes from. And there is nothing truer than math.

But the fixed stories, the simplicity in concepts like $2 + 2 = 4$, that give us a surface level of mathematics open the door to new ideas, new dimensions, new perspectives. If we're only looking at the bottom of a cone, we might see the world in a circular view, but if we turn it on its side, there's a new perspective unlocked. The new dimension is below the surface, and it's something we miss if we only consider the circle.

Of course, there are always hints to these perspectives we hadn't previously considered.

Too often closed mindsets are what keep us from truly growing. We may see the patterns in nature, but acting on those patterns, growing our understanding by seeing things we hadn't before, is difficult. It's often met with such disdain for the negative feelings we feel when we're subjected to them.

But what are these supposedly negative challenges?

To many, they're the mistakes we make. It is easy to be tempted to hide these mistakes from others. This is a high-stakes issue for me because I am the face of an online educational company where how others perceive me matters to the well-being of a large community. I used to think making mistakes would tarnish this perception. But the truth is, I am constantly making mistakes and not getting things "right," so why should I hide the truth?

An unhealthy fear of mistakes leads to a life of misery and one from which we can't recover. We are designed to grow and to create, and sheltering ourselves by avoiding mistakes stunts us. The negative emotions we feel when dealing with the mistakes of our present or past instead can be a stepping stone for the future.

At the core of my basic understanding is my faith. It's a richness of character building I hadn't previously understood, but it's become the basis for my growth. There are depths to understanding myself, my world, and the universe that can't be contained in the pages of a textbook.

My personal growth has come as a result of the resiliency of my faith and the persistence of a God who would always have me come back to Him.

Perhaps one of our greatest weaknesses is our fear of the unknown, but there is always a light pulling us back. There is always a purpose in our movements. There is always a way to return. So, perhaps the fear of the negative isn't quite as real as we might think.

Conclusion

Reaching beneath zero may appear a fool's errand, but it actually solves many questions when we place zero at differing locations. And the theory of negativity has created a landing place for theories regarding the past and even time and space.

But the truly special nature of negative numbers comes from their ability to complete symmetry. The patterns created would not exist had modern mathematics not created a space for them. And, perhaps, creating your own definitions of success and failure may inspire the creation of your own perceptions and pave the way for a brighter future.

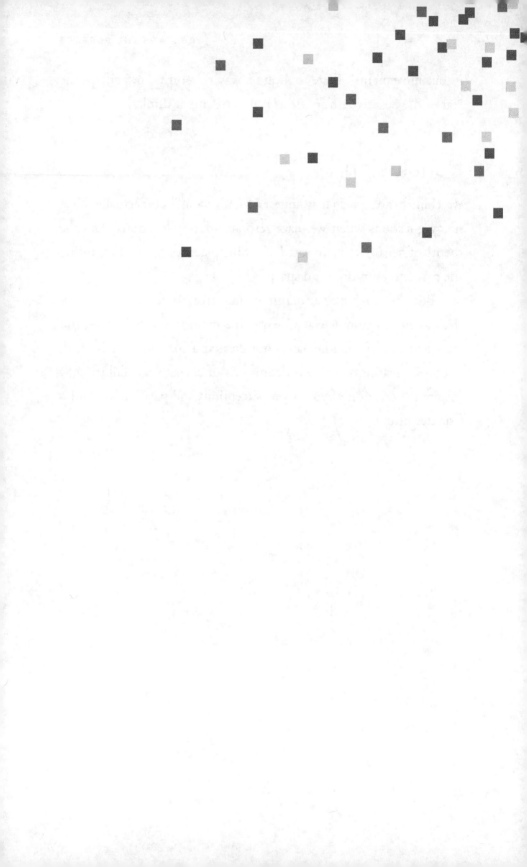

CHAPTER 4

The Fractional Realm

A fractal is a way of seeing infinity.

—BENOIT MANDELBROT, TWENTIETH CENTURY

s there structure in between the lines?

The whole numbers, the numbers not followed by a decimal, hold their own keys to eternity. Infinity moves fluidly in either direction, forward or backward. And at the center lies a definitive equilibrium. The endless array of patterns crafted from single steps forward and backward on the number scale has kept mathematicians enthralled for millennia.

And yet, there is so much more.

The chaos of numbers that lie between zero and one is vast, but what if it's more structured than we realized? How organized can it become?

Rational numbers are essentially parts of a whole, often shown in fractions using integers.

Because a straight line is easy to analyze, we'll start with a foot-long ruler. Looking at the space between zero and one, the biggest division lies directly in the center of the two numbers, denoted by ½. The next-largest ticks correspond with ¼, then ⅛, then ¹⁄₁₆. Unfortunately, at this point, the numbers stop. There are only so many tick marks that fit within one inch. For most purposes, $\frac{1}{16}^{th}$ of an inch is all you'll ever need. If you ever need something smaller than that to retile your kitchen, most people will just wave it off with a shrug: "Close enough."

In the United States, fractions with regard to measuring require much more memorization, and most people just change their units of measurement to avoid confusion. For example, you're much more likely to say you walked a quarter of a mile than 1,320 feet.

But when we move outside the United States, fractions become much more organized: in one kilometer, there are 1,000 meters; in one meter, there are 1,000 millimeters; and in one millimeter, there are 1,000 micrometers. The fractions here play on one another.

Although it's unlikely that you'll ever need something as small as a micrometer when measuring the length of your new TV, it begs the question, how far does it really go down? At what point do we *stop* dividing by integers?

Cracking the Rational Code

Rational numbers began as a way to find more precision in calculations. If you take one step forward, how far is that, really? Does someone half your size have the same stride length as you? (Unfortunately for the poor child required to move with the same stride length you do, they might have to wear stiletto heels to lengthen their legs.)

Much like positive integers, rational numbers came into existence because they're easy to see. They're tangible. Breaking a single great pyramid into the individual stones is a simpler way to look at how everything fit together. Each individual piece—one in 2.3 million stones in the pyramid at Giza—makes a complete pyramid.

And with rational numbers, there are seemingly no gaps left unfilled. Everything is part of a whole. It was another concept easy to grasp. Tangible concepts, especially those relating to geometry, are some of the easiest to justify. And they're the basis for more in-depth mathematical understanding.

RATIONALIZING FRACTIONS

Rational numbers are a natural consequence of developing a system of mathematics. It's a concept never more evident than in the construction of number systems. Integers, as you'll recall, became much easier to manage when set within a base system: base 60 among the Mesopotamians and later base 10 among other Middle Eastern civilizations. Each base system was constructed around the simplicity of breaking large numbers down. It's a grouping we use today as well—the metric system is an excellent example as each unit is separated by a factor of 10, and most countries' monetary systems are divided into hundreds.

Of course, the theories surrounding rational numbers and fractions were hardly useful in early counting systems. Their practical applications became standard when associated with something real. The value of money became quickly associated with the value of goods, and systems of fractions and rational numbers became a centralized way to measure weight.

The weight of one *deben*—a measure of metal, typically copper, bronze, gold, or silver and equivalent to around 91 grams—could purchase certain goods in ancient Egypt. For example, in early texts, one ox sold for 120 *debens* of copper, and two pots of fat sold for 60 *debens*. The type of metal had little meaning in early monetary systems.

It was an ingenious way to keep track of purchases. And, the more advanced civilizations became, the further the divisions went.

On occasion, *deben* weights didn't equate to exact trades. What do you do, for example, if you only wish to sell a single fruit valued at a price less than a single *deben*? The answer was division. *Debens* were reduced further by sharing them or cutting them into fractions. If the weight of copper shavings equaled half of a whole *deben*, the shape of the shavings hardly mattered.[15]

This small dive into rational numbers created a standard that pervaded Egypt. Base rules for dividing numbers into smaller fractions developed into more abstract algebraic problems as a form of necessity: bartering and trading required basic arithmetic. Partitioning the number of bread loaves and beer was an exact science.

But the use of exclusively algebraic rational numbers grew increasingly insufficient as civilizations grew. The different base systems were difficult to navigate as the civilized use of mathematics became more mainstream, making them increasingly more outdated. But perhaps a more uniform method of understanding math would provide a more universal look at numbers.

Again, as mentioned so often before, the Greeks were the first to develop a simplistic approach to fractions. Each simple shape has a pattern of symmetry. It's possible to find the length of all sides of a triangle when you can measure other aspects of these triangles. Measuring the angles and comparing them using formulas can change your perception of the triangle. We can also approximate the perimeter of a rectangle because of the relation of its sides to one another.

In fact, measurement aspects of simple shapes could be related to each other in terms of ratios.

---

15   Although there were many monetary systems that used metal as currency, most of them used different-sized pieces to indicate different values, much like we do today. Egyptians surely weren't the first to think of shaving money into pieces to use fractions, but they were among the first to document it.

Pythagoras remains the most recognized face of this ratio belief. All numbers, he reasoned, could be represented as ratios. Musical notes are denoted by divisions of bar length,[16] recipes divide measurements into segments related to 1, and architectural designs use the ratios of sides to create masterpieces. They're all pieces of a whole, separated into ratios.

The ratios of similar objects weren't the end of the story. If simple shapes have ratios, doesn't it stand to reason that other simple shapes that divide those shapes tell us even more?

It may seem simple to you, versed as you are in modern math, to realize that, within a hexagon, there are a series of triangles, and parallelograms lie within the intersecting corners. Each of these increasingly smaller shapes is intimately related. And analyzing how they all fit together gives us a basis for understanding more about other similar objects.

---

16   For those unfamiliar with musical notation, a bar—sometimes referred to as a measure—represents a time segment in which a certain number of beats is assigned. For example, in a 4/4 time signature, there are four beats within a bar. Each of the notes within the measure is denoted by a division of 1: whole notes, half notes, quarter notes, etc.

Figure 4.1: *An example to consider when thinking about how geometric shapes work with one another. In this diagram, you can see parallelograms, triangles, a pentagon, and more. We can find out more about the shape by analyzing the smaller components. These are hardly the only things you can see within the shapes. How would adding other shapes give us a more accurate look at this object?*

This idea of finding what was inside inspired the name for rational numbers. No, they weren't named that way because they were the next rational conclusion for expanding mathematics. They were named after the geometric pattern of thinking that inspired their creation: ratios.

Though seemingly a simple progression, the addition of rational numbers led to a startling realization: math needed to expand significantly to accommodate the vast increase in numbers. Rational numbers weren't pure. They weren't *natural*. Many cultures until this point had believed that whole numbers would suffice.

The addition of rational numbers also necessitated an increase in intelligible understanding of arithmetic. How, for example, do we know that the whole we think we see before us actually represents a single unit? Isn't it possible that the pizza in front of you is missing a miniscule piece? And, isn't it possible to see the whole pizza as a series of smaller pieces?

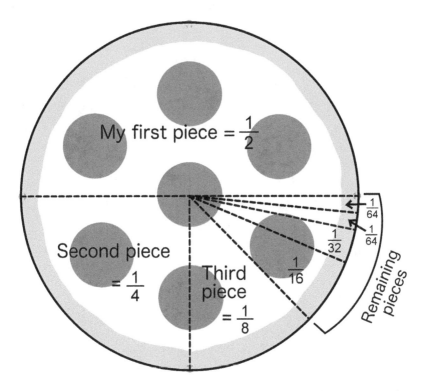

Figure 4.2: *Like a pizza, we can imagine it divided nearly continuously. At some point, you may not even be able to see the smallest pieces.*

The Greeks viewed rational numbers as the unification of all aspects of the Earth. Each piece we see is part of a larger whole.

And what is intimately related to the Earth must also apply to the universe.

The calculated distances of stars from one another, the way the motion of objects in space related to one another all accounted for the inner workings of the universe. Rational numbers depicted a completeness. The sky the Greeks peered at showed not just a jumble of stars but a series of concentric circles that worked together to form larger bodies. The geocentric model—the model that depicted the Earth at the center of the universe—though flawed, illuminated the ratios in space. The universe, as it would appear, held a beautiful pattern.

The changing world that had slowed the evolution of numbers in the past presented an unavoidable slowness in the evolution of numbers. Though ratios and fractions remained floating about the Earth for years, they were primarily used as the helpmates of positive integers in arithmetic.

It wasn't until the mathematician al-Uqlidisi—a name that literally means "the Euclidean"—published a mathematical study depicting rational numbers in decimal form that the fascination for rational numbers grew.

Strange though it may seem, the decimal offered a different perspective to rational numbers. Previous fractions had depended on geometry, the way different lines fit together. However, with the new decimal system, doors opened to more complex math. The addition of zero to the number system had the unexpected perk of providing a placement for more advanced math. It provided a simplicity that made math much more universal. A mathematical problem that might once have been written as $1/10{,}000$ was simplified to 0.0001.

Commerce, scientific discovery, and even the emergence of culture could now reasonably rely on a unified numerical system.

But news travels slowly.

It wasn't until the 1500s that the decimal system became mainstream in Europe. And again, mathematics advanced because of practicality.

Simon Stevin, a mathematician previously best noted for his discovery of the law of the inclined plane and for dropping spheres of lead to analyze the sounds they made when they hit a board, found fractions most difficult to use in bookkeeping. Mathematicians had used the decimal system for centuries, but it hadn't met the common man. And perhaps the secret of its success lay in the snake-oil-salesman-like pitch he brought to those looking for a simple solution to daily calculations. His greatest wish was to teach "how to perform with an ease, unheard of, all computations necessary between men and integers without fractions."[17]

Despite years of crawling back from near oblivion, math simplification instigated the rise of mathematical fashion. The more accessible it is to everyday people, the more likely it is to stick.

The broadened spectrum of usable numbers within the number system caused a chain reaction. Zero had a place in rational number conversation, so why not negative numbers? Who was to say that there was anything wrong with adding an infinite number of tiny divisions between the numbers of 0 and −1? And if that was possible, wasn't it possible that it could potentially extend into either side of infinity?

---

17   It may seem outlandish to pitch simplistic fractional math like a sales pitch, but, in reality, innovation and information spread best when they're pitched as something that will improve lives. He might have tried to oversell it with his simplistic method, but decimal notation is still most commonly attached to his name.

Like those who fought for the acceptance of zero, accepting rational numbers into the number line won another type of battle. It was infinity, really. Only, it was another kind.

THE RATIONAL SPECTRUM

Perhaps the most beautiful results in mathematics are those that, despite their complex nature, just *make sense*. It's like putting together pieces of a puzzle to form a whole. Unfortunately, it's something that most people struggle to find when using math.

If you're like most people who have gone through a math class, you've likely heard the comment "Most people are either good at geometry or algebra; so, if you're not good at one, you'll excel at the other." What a strange thing to tell a child. It's a wonder that so many of us left math with a decent grasp on the subject. If there's anything the rational number journey has taught us, it's that algebra and geometry are inherently connected. In fact, it's difficult to understand the world of math if you don't see their connections to each other.

Geometry provides a visual to algebra's theory. The number line is inherently continuous. It's a property we call the density property: Between any two rational numbers, no matter how close they are to each other, there exists another rational number.

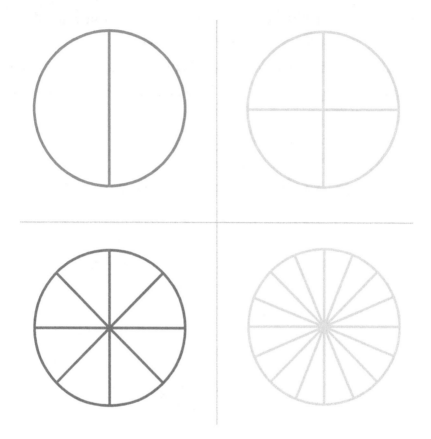

Figure 4.3: *No matter how many times you divide an integer, there is always another number that halves it. This could potentially go on forever. At some point, the thickness of the lines can't visually depict how much we can divide an object, but we can keep dividing.*

We can continue to divide by different integers into infinity, and because there is an infinite number of integers, there must exist an infinite number of rational numbers.

At some point the attempt to reach the smallest fraction becomes like the exercise to reach the number zero. At some point, the numbers become so small that their exact size becomes theoretical. You can't imagine how small the tiniest division is, so how can you fathom reaching the smallest fraction mathematically?

Instead of constantly dividing, what if, instead, we chose to add a series of fractions? Infinite series are a long-form way to answer the question, "What happens if we keep adding to fractions?"

Perhaps what most people immediately consider is the infinite harmonic series. It's the sum of the reciprocals of whole numbers; in other words, $1 + \frac{1}{2} + \frac{1}{3} + \frac{1}{4} + \frac{1}{5} + \frac{1}{6} + \cdots$ adding forever. What conclusion can we draw? The series, like those involved in exponential growth, sees a continual upward trend. But, as you might expect, it tapers off as the fractions become increasingly smaller. It would be easy to assume that, at some point, the sum would approach a whole number. But it never does.

Figure 4.3 is a graph of the harmonic series where each dot represents the result of summing the next term. So, the first dot is 1, the second dot is $1 + \frac{1}{2}$, the third dot is $1 + \frac{1}{2} + \frac{1}{3}$, and this pattern repeats. We connected the dots to form a smooth continuous curve.

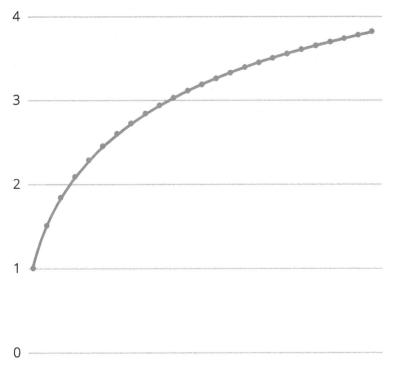

Figure 4.4: *The infinite series depicted here isn't so infinite. It only includes the summations up to 1/25, but it shows a unique pattern. The image rounds off but never completely fades. Since infinity is more of a concept than it is an actual number, we don't know how small we could make a fraction. But, no matter how large the number in the denominator, this curve will never stop growing.*

It's curious. Why wouldn't it reach a finite number?

Well, presumably, if the sum continued to increase forever, there would be no point at which the sum concludes. Just as there is no definitive, observed "zero," there it's impossible to reach a fraction so small that it wouldn't add to the slope's continuous climb.

But the nature of other infinite series behaves quite differently. It's counterintuitive. If the harmonic series continues to grow ad infinitum, surely any other infinite series would as well.

There is a grain of truth in that, which lines us up nicely for another mystery: the geometric series.

The geometric series looks slightly different from the harmonic series. Instead of adding the inverse of whole numbers, we instead multiply the previous number by 1/2. So, the sum of the series instead looks like this: $\frac{1}{2}+\frac{1}{4}+\frac{1}{8}+\frac{1}{16}+\frac{1}{32}+...$ and continually halving for eternity. As you'd imagine, the numbers become significantly smaller increasingly faster. But, instead of showing a steady incline, the sum of the series actually appears to stop just shy of the number 1.

Figure 4.4 is a graph of the sum of the geometric series where each dot is the sum from adding the next term. Again, we connect the dots with a smooth curve.

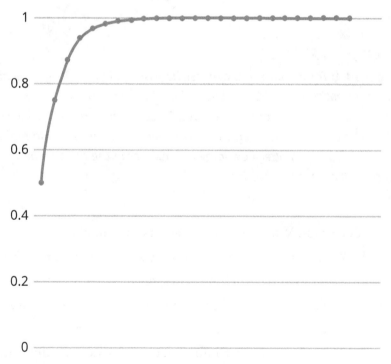

Figure 4.5: *The harmonic series attempts to converge at the number 1 but never quite makes it. It continually grows, though the difference between the points becomes increasingly smaller.*

You may recall, this looks oddly similar to another theoretical concept we've approached earlier in the book: limits. Though perhaps this is a more comfortable look at limits. Because the actual presence of nothingness is impossible, we can instead look at infinity squarely in a zone that most people recognize.

Rational numbers are simplistic in nature. After all, it's always possible to split something. They're part of a whole, so they seem comforting. But if you look closer at the nature of dividing an infinite number of times, it is much more disconcerting. If we split integers, at what point do we reach the end? And is there truly an end to reach?

Uncovering the Rational Continuum

Rational number patterns are uniquely important in turning pieces into a whole. And looking at all the ways we can reassemble rational numbers into a cohesive whole is an important exercise in understanding how rational numbers work together.

But what about the individual numbers?

Rational numbers should make sense; it's why we call them *rational* (alright, it's not, but you get the point). But if we looked at individual rational numbers, their inherent infinite nature is much more surprising.

Since the emergence of decimal notation as a shorthand for fractions, numbers like $1/3$ have become much more difficult to express. In terms of decimals, the number is $0.\overline{3}$, where the bar symbolizes that the number continues on forever. No matter how many 3s we write, we'll never reach a true $1/3$.

This leaves us with an uncomfortable paradox: while the two numbers are mathematically equal, we still have two different decimal representations for the same number, and they appear quite different.

After all, ½ very neatly creates 0.5, and ¼ very nicely follows the trend to create 0.25. How is it, then, that $1/3$ is so elusive?

TAKING RATIONAL TO INFINITY

For millennia, everyone from ordinary folks to genius mathematicians have struggled with the idea of infinity within a simple fraction. Rational numbers were developed as a way to expand math by simplifying measurements, by making sense of difficult concepts. After all, a word is *easy* to pronounce when it's sounded out, a dish is *easy* to create when separated into ingredients, and complex mathematical problems can become easier to solve when they're broken down into essential parts of an equation.

It was a question that plagued the minds of the most mathematically elite. Nearly everything else had an answer. There was always a pattern to uncover. And yet one of the most difficult questions to answer became infuriating when looking up at one of the most common wall decorations: a clock.

The clock depicts the tick marks of sixty minutes within an hour. Within those minutes are sixty seconds. It's no accident—it's a remnant of the early sexagesimal system practices by the Babylonians and Sumerians. The divisions make sense.

But calculating the distance around the circle or the area was another story because both involved the elusive number π.

Because π was a constant accepted since nearly 4,000 years ago, it became a staple in mathematics. But it carried a stigma with it[18] because even though it was a clear-cut fraction of the circumference

---

18 If you'd like to see more with regard to the struggle to accept an irrational number, just wait. The frustration of accepting a constant that wasn't a fraction is a discussion we'll save for the next chapter.

divided by the diameter, 3.14159… wasn't a fraction we could write with only integers.

The question plagued Gottfried Wilhelm Leibniz, a German mathematician, philosopher, and contributor to calculus. Painstaking experimentation and theoretical conjecture had allowed him to approximate the value of π to eleven decimal places. Since π is considered a universal truth, there should be an approximation for it using integer fractions, right? Sure, we have the approximations such as $\frac{22}{7}$ but perhaps there is something more symmetrical to estimate π.

Perhaps unwittingly, Leibniz stumbled across a method for determining $\pi/4$. The method involved creating an equation in which the right-hand side is the sum of an infinite series where the numerators were all 1 and the denominators were positive odd numbers. Instead of the typical summation of all these fractions, the series instead alternated between positive and negative signs, looking something like

$$\frac{\pi}{4} = 1 - \frac{1}{3} + \frac{1}{5} - \frac{1}{7} + \frac{1}{9} - \frac{1}{11} + \cdots .$$

The series is infinite and requires one thousand terms to be accurate to the first three decimal places.

The near impossibility of approximating π by using the sum of an infinite series had instigated proofs for calculus. The process piqued the interest of Leonhard Euler nearly two hundred years later. This time, the number attributed to his name was e, the constant often used to model exponential growth. The number was fundamental to understanding more complex mathematics, but not much was known about the number itself. And, like Leibniz before him, Euler relied on the use of infinite series to ground his explanations of the number.

He'd spent years perfecting infinite series, creating a number of them still used in calculus today. While studying at the Berlin Academy of Sciences, he made a groundbreaking discovery: e could be represented by a sum of an infinite series of fractions using factorials;

3 factorial is $1 \times 2 \times 3$ and is written as $3!$ and $4! = 4 \times 3 \times 2 \times 1$, etc. The final result was written as $e = 0 + \frac{1}{1!} + \frac{1}{2!} + \frac{1}{3!} + \frac{1}{4!} + \cdots$.

It was a revolutionary idea. No one had before considered that e could be written as a sum of an infinite series. Better yet, the sum of the formula revealed a number more accurate than any previous estimations. Certain irrational numbers, then, could be written in the form of rational numbers seen in an infinite perspective. The remarkable idea here is that as long as we limit our sum to a finite number of terms, regardless of how many terms that is, the result will be a rational approximation to e. If we could sum, say, the first one hundred trillion terms, that approximation would be extremely accurate. But it still is a *rational* approximation. So, how do we exit the rational world and enter the world of irrational numbers? We assume our sum continues forever and we introduce the concept of a limit to arrive at an irrational result.

Finding a sum of an infinite series for π, though involving a set pattern of inverted odd whole numbers, requires an infinite amount of them. As the series approaches infinity, the approximation for $\pi/4$ and e only becomes more accurate. Assuming we had the technology to calculate eternity, we could theoretically name all the decimal points associated with π and e. But is it a fool's errand?

The question that continues to plague mathematicians and philosophers alike is the concept of achieving a perfect answer. As it turns out, it is theoretically possible to find all of the decimal points associated with π. But is it realistic? And how far can we dive down into the miniscule before we find the bottom? Or is there a bottom at all?

THE RATIONAL OF TINY REALMS

We've explored the fact that, at the universe's most basic level, it's lumpy. Because there is always a *something*, it's possible to dig deep enough to find the most basic parts of the universe.

The further we go down, the more we see.

Before the 1890s, atoms were believed to be the most basic structure. In fact, the name comes from the Greek word *atomos*, which means "indivisible" or "uncuttable." It was a feat to discover anything smaller than that. Imagining was rudimentary, and experiments to test the behaviors of particles were complex and difficult to perform.

For millennia, it was easier to say that there was nothing smaller than an atom.

But when experimentation and imaging improved dramatically in the twentieth century, that theory was turned on its head. First the electron then the proton and neutron were discovered. And if these particles existed, how many more were there?

True discovery of particle basic movement and properties made scientists ask, what fueled these properties? Could there be something much smaller than originally thought?

At the most basic parts of our universe lie the subatomic particles. We haven't split the electron, but the proton and neutron, at weights roughly two thousand times larger than those of an electron, have particles within them called quarks. There are six in total: up, down, charm, strange, top, and bottom. Each has its own intrinsic weight, mass, color, charge, and spin. Though all the unique aspects of each particle play a part in how they interact with the world, the spin and charge of each particle are perhaps the most curious.

To put it simply, the spin of a particle describes the particle's intrinsic angular momentum. Each particle moves at a consistent speed, at a consistent rhythm. The spin in each particle doesn't refer to a spin around its access but more about its movement in space. Each quark has a spin of ½. The combination of quarks within protons and neutrons determines the angular momentum of the particle. Since both protons and neutrons consist of three quarks—protons contain two quarks with spins of +½ and one with spin –½ while neutrons contain two with spin –½ and one with +½—a combination of the two determines the magnetic properties of an atom.

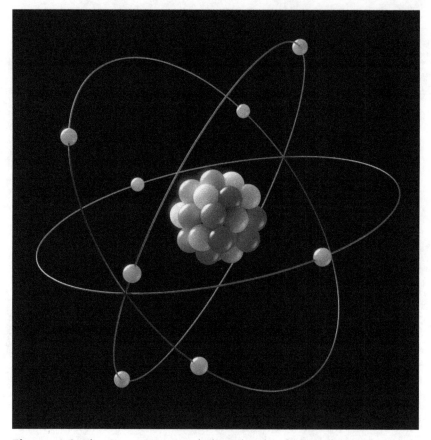

Figure 4.6: *The atom was once believed to be the smallest structure in the universe, but the properties that made up its interactions with other particles were enough to encourage scientists to discover the fundamental makeup of the universe.*

On the other hand, the charge of quarks is divided into thirds. Up quarks have a charge of $+2/3e$ (where e here represents the elementary charge) and down quarks have a charge of $-1/3e$. Protons, again, contain two up quarks and one down quark, totaling a charge of 1; neutrons contain two down quarks and one up quark, totaling a neutral charge.

The information just mentioned—though itself at a very surface level—provides a rudimentary understanding of what happens at a subatomic level. The properties of each particle play a significant role in magnetization, chemical behavior, quantum behavior, and electrical conductivity. In essence, the way that the tiniest particles in the universe react relies entirely on innate properties that are denoted by fractions.

And it's not an accident.

The half-integer spins and third-integer charges affect the way these particles, called fermions, interact with one another. Because of their unique rational spin numbers, quarks cannot occupy the same space as particles with like properties. Other particles like photons—the light particles—have integer spins, and these unique spins allow them to move and interact together.

Rational spin and charge properties in fermions are essential to allow them to transform by their interactions with each other. Without them, atoms couldn't interact with each other the way they do. Molecule reactions would look extremely different, and perhaps not exist at all.

It may be strange to think that, at the subatomic level, one of the most important properties to life exists in a fractional state. It's the combination of a variety of the universe's smallest particles to create a functioning atom.

Though it took so long for us to find the properties associated with the smallest particles in the universe, it opened the door to discoveries of other possible particles including gravitons (the theoretical gravity particles), weakly interacting massive particles (also known as WIMPs and the key components of dark matter), and tachyons (particles that can move faster than light and can potentially move backward in time). What properties could these potentially hold that affect our universe? And can we possibly manipulate their properties to alter the universe?

Living Rationally

Rational numbers are about taking a whole and dividing it into equal parts, like a finely tuned meter stick. Most of my life I followed this rational approach as I pursued success.

I helped build our online educational company by taking actuarial concepts and breaking them down into smaller digestible thoughts. We identified the obstacles to passing actuarial exams and rationally created pragmatic steps for customers to succeed in passing these exams. Of course, when the customers succeed, so does the company and as the owner of the company, I benefit from that success.

Even though this type of success can be easy to measure and has its benefits, I found as I examined my life closer, there were gaps in between the measures of success. Since doing math well in a community is a core value, I wanted to understand these gaps and fill them with something better. The gaps missing from success were virtues more difficult to measure, such as joy and quality of relationships. It's like there were numbers on a meter stick I had not yet uncovered in my journey to success.

It is easy to ignore issues we can't quantify but the last five years or so I have wrestled with this and asked myself challenging questions. What really are these gaps and how can I fill them?

This led to wondering how much I truly cared about people around me. Even though we built a company that puts the customer first, I questioned my underlying motive. Was I helping the customer so they could help me or was it to truly help the customer? Unfortunately, this question applied not only to my math community but to others as well. I concluded through all this struggle for success that there were gaps of an authentic concern for others. I had a rational concern for others but lacked authenticity because it was surface level and lacked depth. So, I set out on a journey to find the hidden gaps of developing authentic relationships.

Like solving good math problems, my journey led to asking questions. These questions appeared small on the surface but had deep roots.

Doing math together requires welcoming diverse opinions.

I analyzed my reaction when these diverse opinions occurred. Even though I was a "team player," deep down there was a part of me that thought I was right. To combat this issue, I made a seemingly small change in my thinking.

When someone shared something different, or perhaps what I considered inferior to my way of thinking, I simply paused. In this pause, I accomplished two simple things. First, in the pause I became aware of the issue. Awareness is always a good start. Second, I withheld judgment as to what was true. In fact, I tried to go a step beyond that and assume that this other perspective was an opportunity for me to see something I normally wouldn't see. This simple pause and reflection paid huge dividends into expanding my train of thought.

Rather than evaluating the "accuracy" of what others said, I tried to value what they said, and this led me to value the person more.

Notice at the root of this relationship gap was challenging myself to have an open mind to different perspectives. Just like a small mistake in math can create a large problem when it adds up, this gap occurred in small doses of seemingly routine conversations. What made this a major problem was not the "importance" of the conversation but the pattern as to how frequently it occurred.

It's true, keeping an open mind and truly listening to others is not a new idea. But the true change that occurred for me was the transition from knowledge to application.

As I experienced these life changes, I realized that this is the same muscle that gets stretched in math about withholding judgment of what is true. Even though I routinely stretch my open-mind muscle in math, I needed to stretch that muscle to my relationships.

One of the most beautiful truths in math is the Pythagorean theorem that states $a^2 + b^2 = c^2$ where a and b and are side lengths of a right triangle and c is the hypotenuse length. This is not intuitively obvious and yet we can prove it is true literally in hundreds of different ways. Doing these types of proofs gives us a road map for setting standards of identifying truth.

I concluded that many things in life I assumed were true didn't measure up to this standard and were, in fact, personal opinion. It doesn't mean what I believed was wrong, but it doesn't mean it was absolutely correct either. It necessitated welcoming perspectives that challenged my beliefs. Ironically, even this iron-clad proof of the Pythagorean theorem depends on an assumption regarding what is true of parallel lines. If we change this assumption, the Pythagorean theorem is no longer true. So, even the truth of the Pythagorean theorem is contingent and not absolute.

This principle has guided me through my life and has prompted me to look upward. The perspectives I miss, God doesn't. It's the gentle reminders in the faces of those who love me that prove to me that math is much more than a set of simple, cold facts. It's never felt more alive.

Conclusion

Can you relate to my story? You may be diligently pursuing what you think is right, but do you sense any hidden gaps? Regardless of our background and belief system, one thing we all recognize is there is beauty in authenticity. Do you desire a deeper reality that perhaps may lie on a different dimension than you are pursuing? Is it possible for there to be a fresh perspective of beauty that is outside our human domain?

If so, you are ready for part 2, where we expand beyond the rational numbers and experience the height and depths of the real numbers. In part 2, we fill the hidden gaps of the rational numbers by adding the irrational numbers and create the real numbers. And for a special bonus, we discover numbers outside the real numbers to expand our imagination.

The rational world, indeed, the rational realm, is the basis for what we know, but it's not the end of the conversation. What we know will ultimately take us further than we ever thought possible. For what do you get if you cross into a realm that can't be measured?

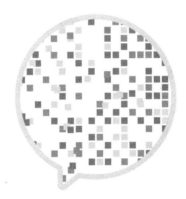

PART 2
Imagining a Better Reality

We can find success with the rational numbers on a calculator, but is that all there is? Are there things in life beyond what we think are rational but worth pursuing? Our rational mind is limited by our perception but is based on the dimensions we observe. But what if there is a greater reality beyond what we view as rational? Can we imagine that there is a more authentic reality that we cannot see from our current perspective? Isn't unconditional love a gift beyond what is rational? I think we all would welcome more love, beauty, and joy in our life. Aren't these gifts we want to give others? I've experienced these gifts but they exist beyond my normal rational thinking and require me to consider a different perspective.

CHAPTER 5
The Irrational Unknown

Isn't it funny how every time we run across an interesting number it turns out to be inexpressible? Maybe numbers like e and π are simply too beautiful to be captured by something as prosaic as a fraction or an algebraic equation. If e were rational, for instance, what numerator and denominator could possibly be good enough? In any case, we have no choice but to simply give names to these things and then incorporate them into our vocabulary.

—PAUL LOCKHART, 2012

What stops us from understanding the impossibilities of math?

Rational numbers, though infinite in their own ways, make sense. They're simply the evenly broken pieces of integers combined to make a whole. If we can't grasp how far we can divide, at least the concept is easy to understand. Even when a loaf of bread is divided into multiple pieces, including the crumbs that fall from the break, combining them together to create a whole is understandable. It's at least somewhat easy to grasp.

But what happens when there is no even piece? What if the division made doesn't result in a tidy fraction? Does that number really exist if we can't quantify it? To view something beyond the rational divisions, it helps to exit the one-dimensional perspective of a number line and consider two and three dimensions.

As it happens, irrational numbers are more commonly found in nature than you might believe. The ratio of a circle's circumference to its diameter—or π—is irrational. The natural logarithm base, the symbol predominantly found in exponential growth—or e—is irrational. They're numbers that are not expressed by fractions using integers. In essence, they're the numbers between the cracks of rational fractions.

But, setting aside the philosophical for a moment, imagine instead a city block.

As a child, playing with friends, you may have become very familiar with the size of a city square. If you lived on one corner and your friend lived on the opposite corner, you probably didn't walk all the way around the block just to see them. At some point, it's just easier jumping fences.

Let's imagine now that you counted your steps walking two sides of the square. At your average pace as a kid, you might have counted 225 steps from your house to the corner and another 225 steps from the corner to your friend's house. Surely taking the shortcut would save you a lot of time, right? But how much distance will you save?

Figure 5.1: *Imagine you wanted to get to your friend's house by jumping some fences. How far would you go if you followed the diagonal line?*

If we use the Pythagorean theorem, the total distance would be $225^2 + 225^2 = c^2$, or $c = 318.19$ feet. You'd save yourself about 182 feet. That's quite a distance!

It turns out that the ratio of one side to the diagonal is $1:\sqrt{2}$. If you're like ancient mathematicians, that might not sit well with you. The square root of 2 can't be expressed as a fraction of integers. So, if it can't be defined by integer division, is it a number at all? The simple number $\sqrt{2}$ is tied to the approximate definition of 1.414, but those are the only two characteristics we can accurately use to define the number. How much do we really trust the irrationality of numbers?

The Irrational Approach

At this point in the book, perhaps it should come as no surprise that the world struggled to accept many numbers we now take for granted. But while the failure to accept some number sets merely slowed down the progress in mathematical development, the uproar created over the addition of irrational numbers completely halted the progress of advanced mathematics for thousands of years.

It's the same concept as reading between the lines. When analyzing a poem, there is a fair bit of leeway in the interpretation. The poem can mean many things to many people. And yet, applying that same logic to numbers doesn't add up.

How do you read between the lines of mathematics?

While there were quantifiable examples of irrational numbers, they simply didn't make sense using the same train of thought as the rational numbers. However, many of these numbers were related to integers and rational numbers through ratios. It was a simple way of looking at special numbers. But surely there must be something, some way, to quantify numbers that doesn't rely on using integers or divisions of integers, right?

Right?

BEYOND PLATO'S INCOMMENSURABLES

The birth of irrational numbers came from a conundrum. Before their inception, speculations of numbers, even zero and negative numbers, came about due to the need for philosophical introspection. But irrational numbers were thrown to the wayside, dismissed as figments of illusion.

An Abstraction Too Far

The Greeks weren't so naïve to believe that all concepts were within their reach. As they continually measured their mathematics with geometry, they found relationships between quantities they couldn't entirely understand. Barriers had already broken in regard to the old ways of thinking. Pythagoras had held an almost reverential idea of whole numbers, believing them to represent characteristics and meanings in life. Nearly anything could be attributed to the ratios of whole numbers, he'd said. The addition of rational numbers, though not strictly conforming to whole numbers, still depended entirely upon the ratios of integers.

Pythagoras banned any talk of irrational numbers in his school. Irrational numbers, or the idea that there were numbers that could exist without manipulating whole numbers, were an abomination, even dangerous.

Ironically, it was by Pythagoras's own hand that irrational numbers were first introduced and it was not on a number line but in two-dimensional space.

When analyzing the ratio of a triangle with two sides equivalent to 1, the hypotenuse must equal $\sqrt{2}$. But, because the number couldn't be expressed in the form of a fraction, Pythagoras simply accepted the existence of a number simply denoted by $\sqrt{2}$. There was a philosophical and aesthetic reason to reject irrationals in Pythagoras's eyes. Anything outside the purity of whole numbers simply didn't exist.[19]

---

19 Pythagoras was quite staunch in his rejection of irrational numbers. According to legend, the ancient Greek philosopher and mathematician Hippasus attempted to solve the problem of irrationality. He discovered that $\sqrt{2}$ couldn't be expressed in a simple fraction in a rudimentary approximation of the number while sailing with his mentor. Pythagoras was so upset by the proof that he threw Hippasus overboard, drowning him. It's important to remember to take the story with a grain of salt, as it is legend. However, it is true that irrational numbers were utterly banned from an entire school of thinking due to Pythagoras.

The refusal to accept irrational numbers was, for the most part, understandable. The concept of decimals didn't exist in ancient Greece. Ratios were the only form used. Instead of writing 1.5, they would express the same quantity as 3:2.

Much of what we understand about numbers today was inexpressible then. When we think of "number" we may think of "1," the algebraic interpretation of the number. The Greeks, on the other hand, viewed these numbers in a far more cohesive sense. The number "1" expressed a unit instead.

Take, for example, the ratio of 3:2 as it's applied to line segments with sizes three and two. The ratio expresses how many units can fit into the different sizes of line segments (in essence, three units could fit within the longer line segment and two units could fit within the smaller). Because both line segments are related to one another by units of measurements, they're considered *commensurable*, or co-measurable.

The most common straw to grasp at is the assumption that any two lines hold some similarities, a link no matter how small the units to express both are. But the Greeks understood from their earliest texts that this wasn't the case.[20]

Irrational numbers aren't expressed in units that compare integers, making them unmeasurable. It was a flaw in written tools.[21] How can you approximate something that isn't measurable? How do you approximate a number that isn't expressible?

---

20 See Appendix chapter 5.

21 Written mathematics the way we see it today is quite an achievement. We often take for granted that we can write math concepts in a way to accommodate more numbers and more math, but it was a hard-won battle. If it's hard to imagine, consider the phrase "You need to see it to believe it." If something is highly conceptual, it's difficult to accept, in many cases. The ancient Greeks were no different.

Because much of ancient Greek math was geometrically based, leaving the hypotenuse of a right triangle presented an isolated solution to a problem. But future generations struggled with the idea that something simply wasn't quantifiable. Was there a solution that could use ratios to express such a difficult concept?

The discipline had been so incredibly steeped in philosophical debate that changes seemed a crime against historical authors of thought. Pythagoras's involvement in the irrational debate was enough, even one hundred years after his death, to dissuade many from approaching it. Even the most prolific writers on mathematics in ancient Greece waffled on the subject; most of the work he produced mentioned irrational numbers as merely an indefinable amount.

Theaetetus, a pupil of Socrates, was among the many who found the perplexing question had an answer. As a child, he'd shown incredible aptitude in mathematics. And yet, it was with its influence that he began to question the methods of number classification. Surely, if there was a $\sqrt{2}$ there also existed square-rooted numbers of other numbers. Legend states that Theaetetus, when asked what he'd learned in school by Socrates, proudly stated that he'd discovered that each number from 3 to 17 also had square roots, and most of them were irrational.[22]

While the means to accurately measure irrational numbers remained, regrettably, out of reach for Theaetetus, he'd created another category now deemed *Plato's incommensurables*. Indefinable though they were, they were at least acknowledged.

The thought experiments Theaetetus may have believed stopped with him actually took flight nearly another one hundred years later.

---

22   We have to give credence to the historians here. Some were so enamored with early Greeks that they might have given Theaetetus a little more credit than he was due as a child. Still, he's credited today with creating another category of numbers.

The change in perspective, the acknowledgment of a more abstract way to look at numbers was appealing to many, though the avenues to reach it seemed nearly impossible.

Euclid, creator of *Elements*, danced around irrational numbers. Though considered a fantasy for many mathematicians, Euclid suggested classifications for these impossible numbers. The locations these square-rooted numbers took on classical geometric shapes and their relations to each other were abstract, yet they held a pattern. In his 115 propositions, his fascination with the diagonal of squares versus rectangles, number theory, and more comprised a fair chunk of his hundreds of pages of writings. What would happen if these numbers did exist? What would they look like if they were given a closer look?

The rectangle and triangles that involved irrational numbers became only the tip of the iceberg. For, where there was one way to manipulate the geometric shapes to create abstract realities with completely different numbers, surely there were others as well.

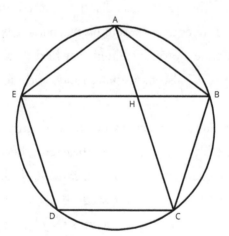

Figure 5.2: *A pentagon within a circle and the triangles formed within it provide the starting point for analyzing the golden ratio. The golden ratio is a natural pattern that exists in nature based on an irrational number. The ancient Greeks used this ratio in many of their works, including architecture.*

The "golden ratio"—a term not used by the ancient Greeks—depicted just how beautiful the patterns developed in analyzing triangles could be. Consider a pentagon with equal sides and angles. When drawing a pentagon, one can also draw a star (or pentagram) inside it by extending the sides of the pentagon until they intersect. This process creates a series of smaller triangles within the pentagon and the star.

The key to finding the golden ratio in this configuration lies in examining the relationships between the lengths of the segments formed by the intersecting lines. Specifically, if you focus on the diagonal of the pentagon (the line connecting two nonadjacent vertices) and compare it to the side of the pentagon, their ratio is the golden ratio (φ). The point where two diagonals intersect divides each diagonal into two segments. The ratio of the length of the longer segment to the shorter segment is also φ.

Another way to see the golden ratio in a pentagon involves the isosceles triangles that can be found within the pentagram. These triangles, often referred to as "golden triangles," have the property that the ratio of the length of one of the equal sides to the base is φ. The angle at the apex of these triangles is typically 36 degrees, contributing further to the pentagon's geometric harmony and its connection to the golden ratio.[23]

The ratio depicts another problem: this golden ratio is also an irrational number equal to approximately 1.618. Because it is an irrational number, it is not a ratio of integers. The commonality of the

---

23   There are several other ratios that prove that this is indeed the "golden ratio." You might be more familiar with particular instances that utilize the golden ratio. Architecture, nature, and even the human body all demonstrate the golden ratio. Leonardo da Vinci is believed to have used the golden ratio as the basis for his famous drawing of the "Vitruvian Man."

number proves that it's not a fluke, that it truly does exist. And yet the Greeks had only integer ratios to depict the peculiar phenomenon.

For years, the Greeks found ways to include irrational numbers within their systems of measurements, even adding them to complex arithmetic and music theory. And yet there was a fundamental disconnect with abstraction. These irrational numbers could be expressed in theory, but numerical representations were out of the question.

Toward the end of the Greeks' reign, the philosophical nature of mathematics began to break down. The term "arithmetic" in Greek essentially translates to "the art of counting," a far cry from what we consider arithmetic to be today.

By the second century CE, one of the final ancient Greek mathematicians, Diophantus, redefined the word to include more complex calculations. More complex equations left room for irrational solutions. However, much like with his predecessors, the solutions that required an irrational solution were discarded, as only *rational* solutions were logical.

Perhaps it was the fear of an infinite progression of numbers that had no measurable conclusion that caused the most fear in the ancient Greeks. There always existed infinity as it related to whole numbers, but the numbers in between seemed an outright contradiction of rational thought.

The disconcerting possibility halted the expansion of irrational number arithmetic for thousands of years. The Greeks created the standard by which many other civilizations based their mathematics, and if they believed that irrational numbers could not exist, it was likely that they couldn't.

For nearly a millennium, the rest of the world slowly picked up the pieces of the fallen Greek empire, shifting from knowledge gained in the Far East to the Middle East. Abstraction of thought became far more acceptable as time progressed, and number abstractions—like those for zero and negative numbers—seemed more plausible.[24] A change in perspective and, even more important, a change in arithmetic style and mathematical notation significantly changed the landscape as philosophical and applicable math meshed into one.

Shortly after the turn of the first millennium, Omar Khayyam incorporated irrational numbers, those still written as square roots, into a new number system. The decimal system, which had developed more or less in India in the sixth or seventh century, made irrational numbers slightly more manageable. But, primarily a poet and astronomer, Khayyam didn't develop the idea much further.

Irrational Enlightenment

Understanding the intricacies of mathematics was a privilege only allowed to the elite. Mathematics had such a power of innovation and advancement that those in charge cracked down on the spread of mathematical and scientific information. Only those in power, the alleged, could handle extensive mathematical theories.

However, like we saw in the last chapter, scientific advancement and its practical development were mostly innovated by average people. Though perhaps relegated to only the basics during the Middle Ages, math was on the rise. Once again, with the resurgence of the decimal system in Europe, mathematicians started to think more abstractly. By the 1500s, irrational numbers became relatively well known. And new numerical notation allowed them space.

---

24   Interestingly enough, when Indian mathematicians came into contact with irrational numbers, they held none of the disdain that their Greek counterparts had for them. They often still regarded square-rooted answers as numbers themselves. The acceptance of this more abstract way of thinking pushed through to the Middle East and made the acceptance of more complex numbers easier.

Proponents of complex mathematics begrudgingly accepted their admittance. A mathematician, Michael Stifel, noted:

> It is rightly disputed whether irrational numbers are true or false. Because in studying geometrical figures, where rational numbers desert us, irrationals take their place, and show precisely what rational numbers are unable to show … We are moved and compelled to admit that they are correct.

It was a natural conclusion to irrational numbers. Simplification of quadratic functions and complex algebraic equations *required* irrational numbers, or at least a definition of them.

Noted scientists during the rise of scientific calculation in Europe in the 1500s came to a rather unique understanding of mathematics, thanks, in part, to irrational numbers. A combination of the geometric model of the Greeks with the arithmetic model adopted by those in the Middle East formed a more cohesive understanding of how an infinite number of whole, rational, negative, and *irrational* numbers could work together in the newly adopted decimal system within European models.

The incorporation of negative numbers and irrational numbers also gave rise to some of the most unique abstractions within mathematics. Previous mathematical attempts to incorporate negative numbers had accepted them as simply theoretical and had left them alone for the most part. Like the Greeks, they were deemed "unsolvable." But with the new introduction of irrational numbers, scientists and mathematicians alike saw the potential of using irrational numbers within negative number systems. Suddenly complex quadratic equations doubled the possible number of answers.

Mathematicians Pierre de Fermat, Blaise Pascal, and René Descartes found the use of irrational numbers and negative numbers a step into the unknown that showed remarkable results when applied in tandem. Irrational powers were considered a "higher transcendental function" as they applied to calculus.

The extension of mathematical concepts to geometry was a natural progression. Descartes, though never explicitly stating his opinion about irrational numbers, baked them into his coordinate system. The accommodation for irrational numbers in basic geometric shapes meant that irrational numbers existed and were graphable.

With the generalized acceptance of irrational numbers came a closer look at what was already a staple in mathematics. Numbers developed thousands of years earlier and accepted as undefinable came under scrutiny. The existence of π and e was well understood. And yet, with the new decimal system, surely it was possible to find approximations for each number.

Johann Heinrich Lambert, a Swiss mathematician, delved deep into the question of π's irrationality. Using the tangent function, Lambert attempted to prove the rationality of π, looking for a contradiction.[25] The consequences of the irrationality of π, far from a mere fascinating piece of information, proved the existence of slivers of numbers between the finely diced whole numbers previously penned.

Once again, there was a new definition for infinity.

IDENTIFYING IRRATIONALITY

Irrational numbers are endlessly continuous and never repeat. It's not hard to grasp why so many people for thousands of years struggled with the concept. An indefinable number can't truly exist, right?

---

25 The rigorous proof of the irrationality of π involves a close look into the tangent function. Following Leibniz's dive into the infinite series of $\pi/4$, he concluded that $\tan(\pi/4)$ cannot be rational, so π itself cannot be rational.

Infinity is a difficult concept to grasp, and it has been for as long as complex mathematics has existed. And yet, without it, we can't fully create a number line. Irrational numbers are the gap-fillers, the in-betweeners that fill up the rest of the number line. Without them, the number line would remain incomplete.

But how fine is the distinction between rational and irrational numbers? Consider, for example, $\sqrt{2}$, which equals 1.4142135624 to ten decimal places (for the sake of this argument, we'll take a tiny chunk of the number). The rational numbers 1.41 and 1.42 border $\sqrt{2}$. If we want to be even more precise, we could say that the rational numbers 1.4142 and 1.4143 border $\sqrt{2}$. The analogy can be taken to nearly infinity. No matter how many decimal places we add to the rational numbers, we'll never find a rational number that equals the irrational.

Though this might seem rather trivial, consider that we can take any integer and divide it rationally into equal parts using division of other integer numbers. Any number can be divided by 10, 100, 1,000, and so on. In fact, the smallest parts of the universe—those defined by Planck's constant—reach only a point of 10^{-34} joule-seconds. It's an impossibility, surely, to reach a number so small that it fits between the cracks.

In the late nineteenth century, Georg Cantor's uncomfortability with infinity led him to analyze points in line segments outside of the realm of rational numbers. Previous generations had accepted the concept of counting to infinity. Though impossible, there was still a theoretical approach to reaching incredibly high numbers, a system that dated back to Aristotle. By the end of the nineteenth century, Cantor had discovered a new branch of mathematics (called the theory of sets) that proved the idea of a "completed infinity."

Cantor dived deeper. He discovered that not only is the irrational set of numbers an infinite set, but it is a larger infinite set than the rational numbers. Being larger does not mean "more of the same." Rather it is a different class of infinity.

To really challenge your thinking, the set of even integers and the set of odd integers and the set of all integers are considered the same size. That doesn't seem possible since the even integers are half of all the integers. But, in the world of infinite sets, they are considered the same size. Even more remarkable, the set of all rational numbers is the same size as the integers. But the irrational numbers are in a different class of infinity that is fundamentally larger.

While uncomfortable, the concept of an infinite number of irrational numbers, a number so large that even an infinite number of integers cannot compare, proves there is no point we can conceive that doesn't exist. We are, it seems, locked into a world of infinite possibilities.

The Mystery of Infinite Irrationality

When we analyze how close we can come to zero without actually touching it, the answer is only possible with abstract concepts; coming so close to reaching nothing is nearly impossible to grasp, and yet it's the only way we can approach it. We could look at any number nearly the same way.

Take, for example, a clock.

Figure 5.3: *A simple clock shows the great connection between rational and irrational numbers. Within the marked divisions lie numbers with infinitely long decimals.*

The clock is divided into twelve parts for twelve hours in the day. Dividing further, each tick on the clock represents a minute, sixty in an hour. While that's where the physical divisions end on a clock, we could potentially take the ticks down further, cramming sixty tiny ticks into the dashes between the minutes. If we added all these tick marks to the clock, the entire outer third might start to look like a black line.

Now, imagine those lines reduced to $\frac{1}{1,000}$ of their current thickness. The significantly smaller lines would create white spaces between the 3,600 black lines. What if we placed a razor-thin toothpick at the center of the clock and spun it? The chances are that the toothpick would point to a location not covered with a black line. What's more, you might not even hit a point that was divisible by sixty at all. You may have just found a new number.

THE IRRATIONALITY CONUNDRUM

Irrational numbers are considered the fragments that fill in the gaps in the number line. After the addition of positive numbers, zero, negative numbers, and rational numbers, it would appear that we've completed the number line. But how true is that? Have we just stopped looking because of the daunting "infinity" label?

It's a question that often comes to mind when considering some of the most famous irrational numbers. In some cases, it's possible to recreate infinite irrational numbers using rational numbers. We've seen that adding a series of fractions leads to an approximation of the irrational numbers. Those series taken to infinity yield much different answers than approximations. And yet, how do we actually know the implications of infinity?

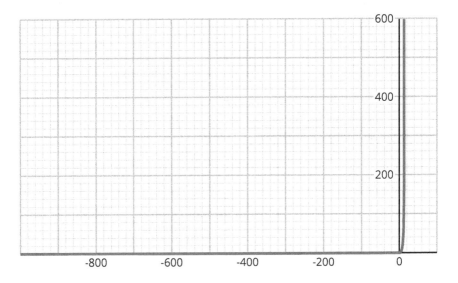

Figure 5.4: *A look at the function* $f(x) = e^x$. *Extremely zoomed out, the function is nearly zero as it approaches negative infinity and doesn't even cross 1 when expanding to positive infinity. Expressing the function without irrational numbers is impossible, as is finding the extremes of both positive and negative infinity.*

Irrational numbers, because they are such a large part of math, often become the staple when discussing the halting progression of mathematics, but they're hardly alone. Just like four of Euclid's five postulates regarding the foundation of geometry remained the standard knowledge, the fifth required thousands of years of analysis and an eventual development of new mathematical frameworks to prove.

The question then becomes, how much is left to prove?

THE INFINITE IRRATIONAL TAPESTRY

With a quick look at nature, you've likely seen patterns that seem unexplainable. The reason salmon jump upriver in mating season, the calculations eagles make when diving to pick up those fish, and the near-instantaneous instinctual fighting techniques these animals make when fighting each other over resources all have a seemingly inexplicable quality to them. And yet, with enough observation, they possess patterns that could predict the way that nature itself reacts.

It's something you've likely discovered yourself, though you might not have known it.

The Animal Base Pattern

When was the last time you could predict the way an animal would behave based on their species? Cats have a hunting instinct that, when unsatisfied, can lead to their tipping mugs over in your home. Cows can chew their cud as a method of more easily digesting food, which must make it through four separate stomachs. Snakes will shed their skin to accommodate growth, repair their injured skin, or to remove parasites. All species have patterns relegated to what they are.

If we dive down to the basest parts of biology, we begin to find that the natural patterns are dictated by one of the smallest parts of our makeup: DNA.

Dogs have an extremely varied DNA base with nearly four hundred breeds. There are potentially limitless combinations of DNA patterns, with a total of $4^{3,000,000,000}$ possible combinations. These traits can predict a dog's levels of anxiety and fear, energy, innate and learned behaviors, and extinction risk.

Figure 5.5: *The variety of dog breeds is based on nearly limitless combinations of DNA traits, each of which is slightly different for each breed. The results are potentially limitless.*

With careful breeding, we can not only predict what kind of behaviors a dog will have, but we can potentially engineer these qualities to fit different situations. Golden retrievers are among the most relaxed of the dog breeds, best suited to be therapy dogs, guide dogs, and family pets. Border collies are ranked as the smartest of the breeds, using their quick-learning abilities to herd, perform agility, and learn tricks. Australian cattle dogs are considered one of the healthiest of breeds, avoiding problems like skin allergies, heart problems, and joint issues.

Of course, because there is such a large variation of gene forma-
tions, it's impossible to engineer a dog that has all those traits, but it's
possible to narrow down the results by analyzing the parents' genes.

Now, let's go further. Environmental factors play a large role
in the development of dog attributes. Behaviors, health, and mental
abilities can all be linked to a dog's socialization, diet, climate, stressors,
training, healthcare, and more.

Truly, a single dog's life is utterly unique.

The Strings That Bind Us

The heart of irrational thought is abstraction, an infinite world created
by something so vast we can't see. It's this logic that has brought about
one of physics' most controversial theories: string theory.

String theory suggests that the universe is made entirely of one-
dimensional strings that propagate through space and interact with
each other. They're made entirely of energy that makes up the fundamental forces created at the start of our universe. It's the result of our
quest for understanding about the birth of the universe and general
relativity to the larger objects.

What would happen, for example, if information falls into a
black hole? According to Samir Mathur, black holes are a "fuzzball" of
strings, and information falling into the swirling pool would simply
redistribute into the fuzzball and emit radiation as it was released.[26]
Leonard Susskind's explanation suggests that the information that falls

---

26   More information can be found at Jennifer Ouellette, "The fuzzball fix for a black
     hole paradox," Quanta Magazine, June 23, 2015, https://www.quantamagazine.org/
     how-fuzzballs-solve-the-black-hole-firewall-paradox-20150623/.

into a black hole is merely reflected off the event horizon, allowing it to escape.[27]

String theory also takes a stab at explaining the complex realm of realities. Instead of our physical three dimensions and a fourth time dimension, string theory suggests the existence of roughly ten dimensions. The presence of dark matter and possible parallel universes all exist under this unproven theory.

The abstract implications of a universe made entirely of strings is possible only through an innate understanding of irrational numbers. Patterns that are unexplainable without an infinite decimal cease to exist. The forces that push matter and cause time to fluctuate are all based on the idea of an infinite, nonrepeating number.

A Superrational Life

My life is a series of patterns, such as my single-minded pattern of pursuing success. Success can be fulfilling but it has gaps in this pattern. I identified hidden patterns of an unhealthy perspective of self. My journey can be summarized by words written long ago by the Apostle Paul. "Do not conform to the pattern of this world, but be transformed by the renewing of your mind."

We do not see transformation by going further but by experiencing change. My solution to these patterns was a renewed mind that resulted from opening my perspective beyond my own understanding. Even though there were positive results from my effort, I couldn't fill the gaps by more effort because effort is bound by my limits. The change that occurred within me was profound, but it resulted

---

27   More information can be found at Tanay Kibe et al., "Black hole complementarity from microstate models: a study of information replication and the encoding in the black hole interior," ArXiv (Cornell University), July 10, 2023, https://doi.org/10.1007/jhep10(2023)096.

from small and subtle approaches. Success can produce happiness and happiness appears to be joy but happiness is different from joy. The difference between happiness and joy is subtle but significant. Happiness depends on circumstance but joy lives outside of circumstance.

Ironically, this transformation from success and happiness to joy is a pattern I see in our number system.

Think of the rational numbers as similar to my pursuit of success and happiness. Through effort, we can produce any rational number on a meter stick simply by slicing and dicing into equal parts. If we want 100 meters, simply take 1 meter, and go end to end one hundred times. If we punch π into our calculator, we may get a number 3.14, which is 3 full meters and then 14 centimeters, or fourteen parts out of one hundred. But is this truly the irrational number π?

It is close to π numerically, but it is not authentic. This method was like my method of being happy by pursuing success. It appears that I have arrived at joy but there is a gap missing. The gap appears small, but it is fundamentally different. We can't fill the gap with more of the same.

So, this leads to the important question of math. How can we arrive at the irrational number π? It's not about going further but it's about finding what is hidden somewhere between 3.14 and 3.15.

Rather than attempting more precise rational division, let's consider a new method that doesn't require slicing and dicing. Consider taking that same meter stick and planting one end in place and then rotating the other end around. This rotation changes our perspective from just looking straight ahead to rotating and seeing things in 360 degrees. The perspective change is like looking at a straight highway and then viewing the Grand Canyon. It is not more of the same but fundamentally different.

So, we answer our question of how to fill the gaps of the rational numbers by this fundamentally different perspective of a circle. Rational numbers can be measured with a one-dimensional distance. But π is fundamentally different because π is not a number on our calculator with more digits after the decimal. Rather π is a relationship built into every circle that connects the distance around to the distance across. Just like our human relationships are packed with intrigue, this relationship that connects π to a circle is packed with intrigue that contains, among other things, infinity, simplicity, and mystery.

It contains infinity because π requires an infinite number of digits to express. It contains simplicity because we generated this circle simply by rotating a stick a full rotation. It contains mystery because it leads to the question of where this relationship originated.

There are a lot of patterns in math. Perhaps you have solved sudoku puzzles. These are popular math patterns, but they are generated from humans. Sudoku patterns are fundamentally different from circle patterns.

Who generated the pattern of π? It is not manufactured by some deep AI logic. The amazing property of the digits of π is that the pattern never repeats. The number appears random and yet it isn't random because it is engineered for a precise purpose.

Because π is not generated by humans, I consider this mystery as a gift that I get the opportunity to unwrap and discover. The remarkable thing about math being a universal language is we can all view the same gift and ask the same questions. These questions are not targeted to one culture, religion, race, or period. These are questions we can ask together and journey into a reality outside ourselves in a way independent of our personal bias.

I'm highlighting π as an example of the wonder of math but π is not the pinnacle of the mystery of numbers but only a drop in the ocean of beautiful math.

For example, the set of irrational numbers is not only an infinite set, but it is a larger infinite set than the set of rational numbers. Hmm. How can something be larger than an infinite set? Notice what this implies is that the gaps we are seeking are larger than the rational numbers that contain the gaps. How can there be more gaps between rational numbers than rational numbers themselves?

Imagine having the ability to create the entire infinite set of rational numbers between 0 and 1 on a meter stick. Realize there are not only gaps between these numbers but more gaps than the rational numbers themselves.

If we consider another simple object, a square with side lengths of 1, we discover a paradox like a circle. On the one hand, a square is a simple shape that appears void of intrigue. But, when we measure the distance across the diagonal, we identify another irrational number, and another infinite mystery appears similar to π.

Where do these simple shapes originate and why do they hide such deep secrets? These mysterious gems are hiding in plain view. I believe that as we ruminate on these perspectives of our real number system, we enter new dimensions and enter a space outside ourselves that is filled with wonder and beauty.

We label these numbers as irrational and perhaps they are from a one-dimensional perspective. But when we enter new dimensions, they transform from being irrational to something beautiful, surprising, and mystifying. There is logic behind these irrational numbers when we view them in the correct context of multiple dimensions. There is nothing irrational about a circle or square. Perhaps a better name is to refer to these infinite numbers as superrational.

Conclusion

The truth is knowing everything as it pertains to something as simple as the number line. The addition of irrational numbers proved that the extent of our knowledge is limited. The knowledge we possess isn't as thorough as we may be taught in school, and it is truly a shame.

We identify π by rotating in a circle. Ironically, this idea of rotating not only introduces π but introduces an entire new class of numbers outside the real number system. We will renew our mind again, this time with a fresh pattern of numbers all resting in a complex plane. What can you imagine we've failed to predict? Better yet, what can you imagine we've failed to find?

The vision of a completely documented mathematical universe is far from reality. There are still elements of space and time that continue to elude us. So, what lies behind what we already know? And what can *you* imagine will change mathematics as we know it?

Sometimes a little imagination, a little impossibility, a little rotation is all it takes to unlock worlds we've never known.

CHAPTER 6

An Imaginary Universe

Imaginary numbers are a fine and wonderful refuge of the divine spirit, almost an amphibian between being and non-being.
—GOTTFRIED LEIBNIZ, SEVENTEENTH CENTURY

Think about creating a work of art on one of the most basic of computer graphics systems: Microsoft Paint. You can push your cursor around the screen, creating dots, lines, basic shapes, and even adding color. Manipulating the pixels on the screen can result in anything from a flagpole to a pretty impressive image of a dog. By the end of your session, you generally have a basic two-dimensional rendering.

Perhaps the closest we can get to true two-dimensionality is on a computer screen. Since the lines you draw there are virtual, they don't have depth. Any effects that make it appear that there is some three-dimensional aspect—such as shading, 3D rendering, or even optical illusions—are merely clever line placements.

However, in real life, there are no truly two-dimensional surfaces.

Imagine putting pen to paper and drawing a single box. To you, that box appears two-dimensional. But if you look closer, you'll see that the indent your pen made into the paper adds depth. In fact, even the ink that your pen deposited has some depth to it, even if it looks infinitely small. When we look at the paper, it may be the lie we tell ourselves to assume that it's two-dimensional, but there are always three.[28]

We often sit on the knowledge of the three dimensions, happy enough with that. It's what we see in terms of length, height, and depth. But there's a little more to it than that, isn't there?

When you see a truck pull an aircraft carrier, when you see a competitive biker struggling to get up a hill, even when you turn the faucet handle to wash your hands, you're using rotation. True, it's still in three dimensions, but rotation isn't something we can depict easily on a Cartesian coordinate system. It requires something a little more flexible.

So, how do you account for that change in motion?

In some science and math disciplines, on the most basic level, we can get away with adding a curling arrow to indicate motion, but it's hardly precise. Since the most basic coordinate system doesn't allow for rotation, is it possible to create a new one?

And does it, in fact, require us to dip a toe into the imaginary?

Navigating the Imaginary Plane

The idea of mapping rotation may seem arbitrary in terms of the number line, but its importance is much more practical. When you look up at the night sky, the snapshot of the universe seems static. But as the year progresses, the stars begin to move. The planets that mingle with them appear to rotate on an axis of their own.

In essence, understanding rotation is a desperate attempt to learn more about our place in the universe. And that struggle was hard-fought.

---

28   If this exercise seems familiar to you, you're not crazy. This same thought experiment was posed in H. G. Wells's *The Time Machine*, providing time as a fourth dimension.

So, more than anything, creating a space for rotation within mathematics required a solid foundation in numbers and graphs. The problem of correctly and accurately depicting graphic rotation isn't new. The Greeks have instilled the need for geometric interpretations of concepts in the foundations of mathematics. And yet, it took until the Renaissance to incorporate them.

BIRTH OF THE IMAGINARY

At the end of the nineteenth century, brothers Ahmed and Mohammed Abd er-Rassul came across an Egyptian burial site that revealed a mathematical papyrus scroll. In it contained the equation for the volume of a pyramid cut parallel to its base, also known as the "frustum" of the pyramid.[29] The find proved what many had already suspected: the Egyptians were among the first to discover complex equations.

The discovery provided an excellent tie to another historical discovery. Heron of Alexandria, a Greek and Egyptian mathematician who lived around the first century CE, manipulated the equation further. The Egyptian equation could find the slant of the pyramid represented by c, concluding in the formula $h = \sqrt{c^2 - 2\left(\frac{a-b}{2}\right)^2}$. One of the first records of Heron's use of the formula—using the values of $a = 28$, $b = 4$, and $c = 15$—resulted in the value $h = \sqrt{81 - 144}$. Unfortunately, he recorded the answer as $h = \sqrt{63}$.

It's unclear if the record was unintentionally changed or not. But, from what we already know about the world's reluctance to accept negative numbers, it's likely that Heron dismissed the possibility of a square-rooted negative.

---

29   The equation for the formula is $V = \frac{1}{3}(a^2 + ab + b^2)$, and it's quite an impressive mathematical formula for ancient Egypt. The formula works for every pyramid using only geometry and trigonometry.

Two centuries later, the Greek mathematician Diophantus[30] also dodged the issue, instead keeping particularly disagreeable results in squared format. Because there could be nothing less than nothing, it was simply easier to leave it that way.

It would take over a millennium before a square-rooted negative number would find its way back into mathematics, this time in western Europe.

There is a certain level of mystery to imaginary number origins. Most of the acclaim for the first acknowledgment of their existence revolves around the controversial bout between mathematicians Girolamo Cardano and Niccolò Tartaglia. Both claim to have played a part in the widespread acceptance of the square root of a negative number. Both wrote books containing possible mathematical solutions with the concept. And yet, it's likely that both received inspiration from other sources.[31]

Still, it was their contemporary, Rafael Bombelli, who officially defined them in terms of algebraic calculations. It was the first time the words "The square root of minus one is equal to i" had ever been uttered. And yet, it created a philosophical stir. Could the square root of a negative number truly exist in reality?

Imaginary numbers required a look beyond the typical planes of existence. Imaginary numbers represented movement in a way most had never considered.

---

30 Diophantus is considered one of the greatest minds in mathematics and was one of the final Greek mathematicians before changes in Greek political structures and wars caused the practices to decline. His book *Arithmetica* is considered one of the founding books in mathematics, on par with Euclid's *Elements*.

31 The mathematician Scipione del Ferro taught both mathematicians and revolutionized methods of solving cubic math problems. Though his name is mostly lost to history, it's likely his tenure as a professor at Bologna influenced many mathematical discoveries in the 1500s. Secrets were among the most valuable currency of the day, and having knowledge of unsolvable problems made anyone who held those secrets particularly powerful.

Leonard Euler, a Swiss mathematician, envisioned a combination of real numbers and imaginary numbers on the unit circle. Instead of relying on the static numbers in a typical coordinate system, the movement and consistency of sines and cosines—the basics for graphing waves—could possibly provide a better system for breaking down imaginary numbers. And, in fact, they did. By expanding e into an infinite series and by analyzing the rate of change of these functions, he developed one of the most important formulas in history, known as the Euler Identity: $e^{\pm ix} = \cos(x) \pm i\sin(x)$.

To those who aren't math buffs, never fear. In essence, this equation shows us the connection between trigonometry and complex numbers. A complex number is the sum of a real number and an imaginary number. So, $1+2i$ is a complex number as is $\cos(x) + i\sin(x)$. It was revolutionary. Much like the scientists of old, the need to explain complex concepts in terms that were easier to understand, in this case, created a solution no one expected.

And yet, the theoretical aspects of imaginary numbers were still difficult to grasp graphically. The truth was (and still is today for the most part) that the impossibility of an imaginary number was directly linked to its lack of graphability. So, what would $\sqrt{-1}$ look like?

Nothing before the seventeenth century could predict how something that didn't exist square rooted would look on a graph. The impossibility explored by frustrated scientists attempted to find solutions with the combination of basic shapes, lines, and even plotting on the newly formed Cartesian coordinate system, but to no avail. How *do* you graph a figment of the imagination?

Geometrically proving a paradox often comes easiest at the most basic level. In the mid-seventeenth century, John Wallis approached imaginary numbers by manipulating triangles.[32] His constructions yielded a paradox using the Pythagorean theorem. Again, he'd discovered a problem with square rooting a negative answer. If there were no such things as imaginary numbers, why did they show up in basic geometric problems?

Remarkably, the answer came nearly one hundred years later from a cartographer. Caspar Wessel's[33] frustration with vectors led him to a detailed dive into current coordinate systems. We can calculate points on a coordinate system with the Cartesian system, so it's only logical to assume we could do the same for complex numbers. But the concept takes a little bit of imagination. Like a typical *xy*-coordinate system, the horizontal axis still represents the real part of a complex number while the vertical axis could represent the imaginary.

---

32 John Wallis's work attempted to investigate an ambiguous geometric problem: construct a triangle with two unknown side lengths and an angle of unknown size. When analyzing the sides and the angles of this triangle, he found two separate solutions, one resulting in an imaginary number. Though he couldn't explain the result, he suggested the need for a new plane that would accommodate this paradox. It was the foundational experiment for creating a complex plane.

33 Caspar Wessel is among the most inspirational of characters in the voice of math. Not only did he create a new system of thought, but he did so when he was deep into his adulthood (52). And it was created by a man plagued by rheumatism and forced to quit his position as a cartographer. He was also recognized as the first person outside of the Royal Danish Academy of Sciences to receive publication in the Academy's science journal due to the precision of his conclusions. It's true that you can make discoveries at any age.

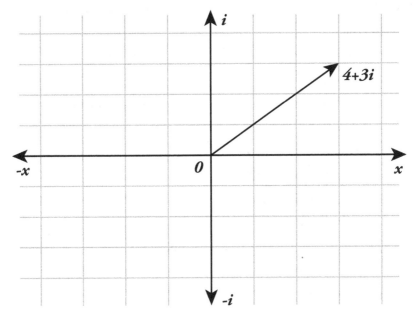

Figure 6.1: *Example of a complex plane with corresponding vector.*

For the first time in history, imaginary and complex numbers were represented in something besides the theoretical.

A LEAP BEYOND THE REAL

Imaginary numbers are among the most difficult to grasp for people entering mathematical spaces, at least at first. Even the greatest minds in history struggled to understand the consequences of imaginary numbers. It's possible to imagine all other numbers within the number line, yet imaginary numbers are a category of their own.

Perhaps it should come as no surprise, then, that imaginary numbers were created as a means to an end. After negative and irrational numbers were added to the number line, it only made sense that they followed the same rules as positive numbers; a square-rooted positive number yielded an integer, fraction, or irrational number.

Though the results may have looked messy, they were still in the realm of reality.

However, when applying the same rules to negative numbers, things get a little trickier. "Fake" numbers like negative numbers (really, numbers that didn't follow the classical understanding of numbers) built on other "fake" numbers created quite a conundrum. And, in fact, it required a completely new type of thinking to envision.

Little did previous mathematicians know that these new numbers would have incredible consequences. Patterns that appear in nature, never seen or even known, emerged after they took a closer look.

One of the simplest forms of imaginary numbers in real time is the humble snowflake. Looking at it from far away, it may just look like a puff of white. Yet, the closer you look, the more complexity you see. Each arm of the snowflake is divided into smaller and smaller pieces, all of which are the result of the freezing process high in the atmosphere.

Figure 6.2: *Imagine the intricacies of a simple snowflake. Though not all of the repeated patterns are of the same basic shapes, the zoomed-in shapes are never overly complex. The second image is a Mandelbrot fractal, dividing each line segment into smaller shapes.*

The snowflake is an example of a fractal. Fractals are mathematical shapes that preserve their shape regardless of how much you zoom in or zoom out.

Now, let's use complex numbers as our canvas. Advanced mathematics using complex numbers allows us to create fractals that are infinitely more intricate. Among the most famous are Mandelbrot fractals.[34] To achieve this repeating pattern, we perform a complex calculation ad infinitum. Depending on the type of complex equation we input, the Mandelbrot could stop after a few repetitions or repeat forever.

This unique pattern also extends into music. A complex frequency, like those found in musical compositions, is a series of frequencies blended together. With mathematical manipulation, we can identify sound waves and the frequencies at which they vibrate. We can use a mathematical prism to split music into pieces, identifying each color in the composition.

The fascinating realization scientists made about complex numbers wasn't on a linear plane. Instead, it required a slight rotation, a slight turn to the side, to recognize an infinite number of new possibilities. The realms of math are infinitely more complex outside of our typical coordinate system. Patterns discovered in complex formulas have led to a better understanding of things we can't see at the surface.

The question becomes, how can we work with numbers outside our field of vision? How does it all work? Imaginary numbers seem to have a magical quality to them, but how far does that extend?

---

34 The Mandelbrot image is broken down into a visual representation of a complex number. To create the image, we square the number and add the original number to the result, repeating an infinite number of times. The points that move closer to infinity are colored with warmer colors, like red, and those that take off at a slower pace are colored in cooler colors, like blue.

Beyond the Complex Plane

Take a moment and return to your favorite bakery. The baker is filling a bowl full of cinnamon and sugar to roll out the perfect snickerdoodle cookie. In the mixing bowl, the baker adds the flour, the sugar, the butter. As one of the final ingredients, the baker dusts in a teaspoon of baking powder.

The dive into baking may seem a stretch to a mathematical conversation, but it's a great analogy for understanding more about the complex plane. The most prominent ingredients in mathematics are e and π, symbolizing the flour, sugar, and butter. But one of the most important ingredients in this cookie analogy is the baking powder. Without it, the cookies wouldn't rise.

Imaginary numbers are the hidden factors in many mathematical and science disciplines, and without them, many wouldn't work.

So, what lies beyond?

IMAGINARY OR NOT?

The real number line is complete with a conspicuous lack of imaginary numbers. And yet, this isn't a flaw. Because complex numbers perform their magic with rotation, they are not needed in our linear vision. To model rotation, we benefit from a completely new number line. The worst part is that depictions in a standard plane don't work. A standard calculator can't compute imaginary numbers graphically. The images are two-dimensional and inherently inferior when depicting rotation. You could say that we can't see complex numbers in the real world.

So, how do we envision them in reality?

The key is to use a different perspective entirely.

Imagine a circle. If we break down the angles we could create or the degrees we could break down within the unit circle, the result will appear like a black blur. But, if we instead start thinking of the circle in terms of its functions instead of the numbers that define individual points, we start to better understand how each section works together.

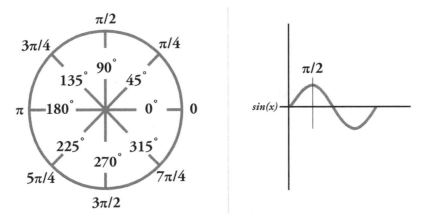

Figure 6.3: *It's the unit circle you've seen a million times, but what if we looked at it from a different perspective? The image shows the circle in terms of degrees and π. But, what we can see in the unit circle can also translate to an extended frequency. Laid end by end, we can see a sine or cosine wave.*

What if, instead of seeing the unit circle as degrees, we saw it in terms of π? Each part of the unit circle becomes a function of a single unit. We can start to see the unit circle as a division of a single function.

Now, let's take it a step further. We can also see the unit circle in terms of sines and cosines, or in the form of a wave. If we were to draw the unit circle but someone pulled the sheet of paper as you drew, you might get something that looks awfully similar to a sine or cosine wave. The two different perspectives tell us something: it's often when we change our perceptions that we see a new avenue, something we might not have seen if we'd stayed in a static position.

The Greeks saw the numbers they used as not simple numbers but as movable tools to shape their perceptions. So, how much more could we change if we zoomed in and out, perhaps manipulating the frequencies of sound around us? How much are we missing by looking at straight lines?

Complex Reality

Coaching Actuaries creates actuarial content for our customers to consume. This content includes written online manuals and videos. Creating this content well requires a balance of science and art. The bottom line is that it is all creative work and in order to have a culture where working together is healthy, there must be a keen awareness for human issues, especially issues that are hidden. I'm thankful for our staff who have worked through those hidden issues, occasionally despite me.

Of course, fostering a creative culture is nothing new. A good resource to see how this has been done well is the book *Creativity, Inc.*, a book about Pixar. Here's an insightful summary quote:

> What makes Pixar special is that we acknowledge we will always have problems, many of them hidden from our view; that we work hard to uncover these problems, even if doing so means making ourselves uncomfortable; and that, when we come across a problem, we marshal all of our energies to solve it. This, more than any elaborate party or turreted workstation, is why I love coming to work in the morning. It is what motivates me and gives me a definite sense of mission.

We can develop this muscle for respecting the hidden by doing math. Doing math well requires a muscle to value things that are hidden.

It's like a nonstop game of hide-and-seek. Sometimes this process is frustrating and can even lead to despair. But it is developing important muscles of having a healthy respect for what we do not know.

One advantage of studying the history of math is seeing the drama unfold. The story of the imaginary number—and complex numbers in general—is a great example. This amazing idea has always existed, but it wasn't found until relatively late in the history of math. Many of the greatest minds of math struggled to identify what it is and how it can be used. We are recipients of that struggle, as the fundamentals of complex numbers are no longer hidden.

I hope you have enjoyed learning about the people and culture that brought numbers to where they are today. It is one of the greatest stories we can share. It is an unending pattern of unveiling what was hidden. These fascinating stories are not stories of independence and isolation, but of building on past ideas, as all of it is somehow connected. I don't want to label it as a happy ending because the journey of discovering patterns of numbers will never end, which is a good thing.

Perhaps before reading this chapter, you had little exposure to the imaginary number. If so, I'm sure this chapter was a challenge to wrestle with how "real" this imaginary number is. Don't dismiss that struggle and just accept it is true. Rather, let your mind struggle and ask questions that can yield a deeper level of understanding. Consider the problems in the Appendix as an opportunity to put this idea into action.

As we reflect on the complete number system there is a wonderful combination of truth and beauty. We receive all the pragmatic numbers we need to measure using the rational numbers. But, hidden on a simple number line is the gift of numbers that speak of infinity with their never-ending digits floating through this math reality where no pattern to these digits exists. It's the absence of a pattern that makes them special. We know within the real number system there exists a number, call it x, where $x \times x = 2$. This number x is precise. Just one digit off and the result of the equation does not equal 2. Yet we know the value of $\sqrt{2}$ is a number with an infinite number of digits that never end in a pattern that does not repeat. What a gift that is!

Then we're given the imaginary number i which produces the complex numbers. So much can be said of the beauty of the complex numbers, which includes the real numbers.

As you explore the set of complex numbers, you may sense a truth that exists beyond yourself. The more you play in the complex plane, the more you realize how this space somehow models deep realities of this world. The amazing mystery is that we extended our x/y-coordinate system to the complex plane by introducing a single new definition that originates in algebra. This certainly doesn't seem logical. We can't explain why this occurs. At some point, we can simply smile and say softly, "Why should we be surprised? It is math."

So many of the good things in math are "hidden." Hidden jewels can be the history of how math has arrived, with the imaginary number a classic example. But they are also the beauty of the math itself. The difference between rational numbers and irrational numbers truly is fascinating, yet it is hidden because the numbers appear the same on the calculator. The imaginary number began as a dismissive step to an algebra question but has been developed into an amazing tool to explain abstract and applied math concepts. What motivates me to write this book is the opportunity to reveal a small sampling of these hidden gems that have impacted my worldview.

What is hidden is often not what we'd expect, and it's frankly how I've become who I am today.

Conclusion

You, the one reading this book, have the skills to be great. We often associate greatness with people who succeeded in memorizing what other people have known, and it's easy to get caught in the belief that you need to do the same.

But the most innovative people in the past blew past that stereotype.

The people who paved the way for more imagination were the people with a different perspective, and I can guarantee, if you've never felt connected to math in the past, you likely have a perspective we need to see more of. We don't just need more memorization of material; we need people who can see patterns. We need people who can think creatively. We need *you*.

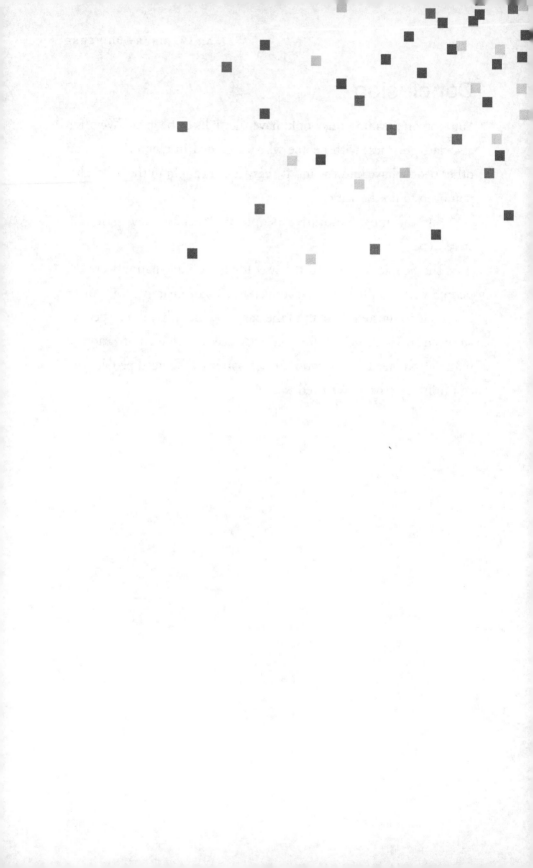

The Transcendental in You

I feel so small

With my hands up to the sky

I am reaching out tonight

'Cause this is bigger than us

I give my all

But it's just too much to hope

No I can't do this alone

'Cause this is bigger than us

Bigger than us

—LYRICS FROM JOSH GROBAN, "BIGGER THAN US"

W e've traveled far on our journey. We arrived at the complete set of real numbers and then we expanded the real numbers to a larger set called the complex numbers. In our last chapter, we briefly visited a subset of the real numbers we refer to as the transcendental numbers. The name given to this set is interesting.

A common dictionary definition of transcendental is "relating to a spiritual or nonphysical realm." To get a basic understanding of what these numbers are, we need a quick flashback to high school algebra.

Recall the drill: given _____, solve for ____. For example, given the equation $x - 1 = 5$ and plugging 6 into the equation, we find that the equation holds true. Likewise, if we have the equation $y^2 - 1 = 0$, we can say 1 and –1 are solutions to these simple equations.

There are certain numbers that will never be a solution to these types of basic equations, and these are what we refer to as the transcendental numbers. It's like these numbers transcend or are too special to live in simple equations. These numbers are real numbers that exist on our number line, but to find them, we must practice the art of looking beyond the one-dimensional space of algebra. It's like these numbers live in the wild and we must travel out of pragmatic math and into an infinite state to identify these numbers. If we only view these numbers as another button on our calculator, it's like taming wild animals in the zoo and we miss their natural characteristics.

We've uncovered three transcendental numbers so far, π, e, and the golden ratio. Some of these numbers we scratch our head and ask where they really came from. Sure, we know π is the ratio of the distance around a circle divided by the distance across, but why and who decided circles would have this remarkable property?

Beyond Measure

It is refreshing to be taken to a place outside us. Sometimes we enter that space by gazing at the night sky and observing the stars above. We wonder how far away these stars are, how many of them are there, and where they originated.

Even though these are amazing concepts, and the answers are beyond what we can measure, they still are finite. The stars are a long distance away, but it is a finite distance. Where can we travel to gaze upon the infinite?

With a little imagination, the surprising answer is a circle. We don't view the infinite from any real circle we create. But we can gaze at the infinite by imagining a transcendental circle created with an exact radius of 1. Then, in this transcendental world, we can see a fingerprint of the infinite by the relationship between the distance around this circle and the distance across. This gaze into the transcendental infinite has been hiding in plain sight your entire life.

Perhaps this gaze stirs a question of curiosity, such as "How do we know there are an infinite number of digits to π?" By "knowing," we don't say we think we know, but we know with 100 percent certainty. How can we be so certain that an infinite concept truly exists?

Principles to Consider

Throughout this book, you've seen how others have struggled to come to terms with new concepts. You've seen what we know about these concepts, some elusive, some not. You've seen how, when taken to the extreme, these numbers provide the springboard from which we can discover some of the mysteries of our universe. But, more than anything, I hope you've seen how these numbers can mirror your life.

I hope there are principles that we've discussed that will transcend the pages or voice of this book.

BE CURIOUS

Perhaps your math experience is someone else gives you questions, and your job is to answer the questions. A fundamental assumption in this book is everyone can do math. Doing math is more about being curious and asking questions than it is about calculating an answer. If you need a place to start, work through the problems in the Appendix and ask questions along the way. The Appendix is designed to spur your curiosity. Another fun way is to use a tool like Desmos and play around with it. Draw a circle. Draw a triangle where all endpoints are on a circle. Write an equation of a line through the center of the circle.

BE LOST

Being lost in math can be a wonderful space to enter. Math is difficult and it humbles us. There really are seemingly crazy ideas in math that appear out of nowhere and don't make sense initially. Be comfortable in that place of uncertainty. Don't rush jumping to the answer. There is tension in not knowing and that tension often is what we need to truly journey forward. Like a tightrope walker needing tension in the rope, we need to embrace the tension of not knowing and consider this as an important muscle that must be stretched.

One benefit to embracing being lost is we are more open to changing direction. We live in a world that celebrates big movements whether it is a miracle or Super Bowl win or a rock star celebrity. It's a win-the-lottery mentality that we hope for. Some of the most important events in our lives can occur from small things but they usually are elements of change, not more of the same. The multiple dimensions of math invite a lifestyle of change and change originates with a willingness to be lost.

CELEBRATE THE FINDING

Hopefully through the stories of this book, you have a deeper appreciation for the struggle of how math has arrived. If you think about it, Archimedes never knew the algebra we know today. Learning subjects like algebra is difficult but learning how to abstract concepts is a tremendous growth experience. So, when you understand a new concept, take a moment and appreciate it. You may have just learned something many of the great mathematicians of the past never knew.

One of the blessings of math is that once we arrive at a sound understanding of a math concept, our brain has an uncanny ability to compress this knowledge so that it requires only a small part of storage space in our brain, and we can retrieve this information quickly. This is really a cool way our brain works but we can quickly forget the effort it required to gain this understanding. You likely already experience this compression phenomena in math concepts you understand well such as addition. But compressing this knowledge can lead us into thinking that this knowledge is now "easy." I'm amazed how "easy" a concept can be in math until I remember that only a few days ago, I thought it was "difficult."

TEACH OTHERS

I've had the opportunity to answer a lot of students' questions and many of these questions have challenged my thinking. I'm amazed as to how much I learn from the process. Not only is it a good source of questions, but I often benefit by explaining concepts to others. There is something almost magical in the teaching process that produces a deeper understanding of a concept. Sharing what we've learned with others can also be a way to celebrate what we've learned.

READ BOOKS

This book can be just the first step of a life-changing journey. My change in perspective of math began when I started reading books about math. The first book I read like this was in February 2017 and it literally changed my life because it created a new pattern of reading different perspectives of math. Many of the books I've read since then are historical books, but some are just detailed books on topics I find interesting. Math textbooks are wonderful, but there are many interesting books that explore different dimensions of math through time, people, and culture.

Reading these types of math brings context to the core ideas. It also illustrates the width and depth of humanity. We have been given wonderful gifts in math, and reading stories of how we humans have discovered these gifts is a shared humanity experience. Math truly is the universal language.

REMEMBER, YOU'RE NOT ALONE

We've highlighted how broad math is: its heights and depths. It's like understanding how big our universe is. On the one hand, it is a source of wonder. But on the other hand, it can be overwhelming, and it is easy to feel lost.

When you take a step back to look at all the patterns in the world, one of the most important is the Voice that orchestrates it. The truth is, in this realm of infinite patterns, the common theme is the existence of an Engineer, a Designer, a Voice that directs it all.

The most important discoveries about math have proven that these patterns aren't here by happenstance.

Though I've gone through various stages of my faith, I've found a divine connection to the eternal through my journey with math. It's a never-ending bliss that keeps giving. And, if you follow the voice of math, you'll find it, too.

EMBRACE MATH'S SELFLESSNESS

Engaging with math encompasses a lot of things, but at its heart, it's selflessness. Though we might have our own notions of truth and right, math can help you recognize that subjective perspectives don't have a handle on math. It is truly the essence of basic truth. It's not contingent on your social status or level of influence or even your intelligence. It simply exists as a pillar of truth.

Math has the unique talent of stripping away the pride we feel. Instead, the joy felt in the pursuit of discovery has encouraged billions to seek math's selflessness for themselves.

FIND AN INVITING SPACE

I enjoy sports and choose to be an active participant my entire life because it not only promotes physical health but is a safe and fair arena to teach me important life lessons. I've benefited from engaging in sports because it is a space to develop discipline, accept defeat, play well with a team, and celebrate the thrill of accomplishment.

Likewise, I actively participate in math because I view math as a safe and fair arena to stretch my understanding of truth, embrace struggle, expand my mind into surprising dimensions, and enjoy the process of finding unexpected relationships.

Math is not likely our end game in life, but it can be an opportunity to enter into a space that has the capacity to change our life and allow it to flourish. I've created a website www.intersectingus.com that is an inviting space to see your math flourish.

With Gratitude and Wonder

I'm thankful for my journey through math and an opportunity to share some of that journey with you. I hope this book serves as a reminder that, with all the confusion in the world, you can find solace in embracing truth. It's the key to human prosperity.

The voice of math is the story of ourselves, of our Designer, and of discovery. We'll forever look for the mysteries that surround us, and

I pray that you'll find the joy in the journey and peace that exceeds your understanding.

Of course, this book is not π so it needs a last digit, a last piece. Quite fitting, the last piece came to me as a surprise gift. All the pieces in this book are a gift to me so this last gift fits the pattern. The very day I was writing this ending, I received a message of encouragement from my friend Brian. Brian wasn't trying to finish my book but was giving me his usual words of encouragement of a hope we both share. In this book, we've journeyed through ancient stories of math that have built the foundation of the math we know today. Let's complete our journey with a message from long ago written by King David in a song of wonder and thanksgiving.

"Great is our Lord, and abundant in power; his understanding is beyond measure."

Additional Practice and Stories

The goal of these stories is to consider some fresh perspectives and think for yourself. The stories are connected to the topic in each chapter. You can choose any of the stories you wish but some topics do build on one another. While this is a great resource to answer some of your questions regarding math, it's merely a stepping stone to help you activate your brain. None of this is graded so enjoy!

Introduction

In the introduction, we explored John Napier's creation of the logarithm and how it converted multiplication to addition. Let's dig into some of the math of this wonderful discovery.

It all begins by considering what thoughts come to mind from the number 3. Upon first thought, perhaps you considered three cats or three children or even three little pigs. What could be simpler than the number 3? Visualizing three dollars is easy—just imagine three one-dollar bills.

However, there are myriad ways to reach that same amount. For instance, we could use twelve quarters, thirty dimes, or even three hundred pennies. We could also combine them, creating three dollars with four quarters, ten dimes, and one hundred pennies.

But for simplicity, let's consider three identical squares. If the length of each side is 1, the area for each square is 1 x 1—or 1—and the total area of all three squares is 3.

We've already considered the number 3 in three unique ways. The most intuitive way to envision 3 is by counting individual items. It's an integer, a number that naturally follows 1 and 2.

We can also conceive of 3 as a combination of smaller parts, all collectively adding up to 3. We use this method so many times when dealing with money it's become second nature to us. Each division of the number 3 is equal and interchangeable.

Finally, we transitioned from simple counting and accumulation to a more abstract concept: measurement. For example, stacking three squares on top of one another creates a two-dimensional area.

Consider this: when we arrange three squares together, we form a rectangle with dimensions of 1 x 3. We've also inadvertently discovered a fundamental formula for calculating the area of a rectangle: length times height. This formula serves as a foundational building block upon which more complex mathematical concepts are constructed, much like our three stacked squares. These varied perspectives and evolving ideas embody the dynamic voices of mathematics.

However, we've only scratched the surface. Until now, we've dealt exclusively with straight lines and angles. Yet, the allure of mathematics often lies in its curves. Let's be honest; curves often add an element of intrigue and beauty to many aspects of life. They also add depth and complexity to mathematics, presenting challenges that make the journey even more rewarding. Let's continue exploring the number 3, but this time, let's delve into its curved interpretations.

Before we embark on this journey into curves, math comes with a word of caution. It's like the warning label on your hot beverage, urging you to sip slowly and carefully.

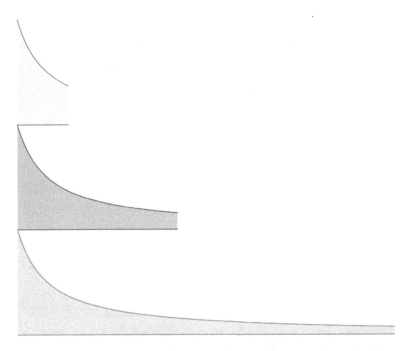

Let's start with three different shapes illustrated in the diagram above. Each has straight lines forming the sides and base, but their tops are defined by a curve. If we compare areas, which has the largest area? Which has the smallest area? Notice that we don't require a unit of measurement here.

That's the beauty of mathematics; it isn't confined by specifics like feet or meters. Instead, we can delve directly into the core: the numbers and concepts themselves. Even though it is common not to consider units in math, that doesn't mean we can't define a standard. To establish a basis for comparison, let's reintroduce one of our squares. Also, to help with our comparison, let's arrange our three shapes from left to right and put our standard square at the far left.

This perspective sheds new light. First, the height of the square is the starting height for the first curve. Since we previously identified the square as 1 unit, the height of the square is one unit. Now that we have our unit of comparison, how would you rank the areas of these three shapes?

Surprisingly, all three curved shapes have the same area. They also have the same area as the square. Remember our square had an area of 1. That means we can say the area for each of these curved shapes is also 1. So, the combined area for all three curved shapes is also 3, providing us with yet another visual representation of the number 3.

Now, let's dig deeper with some simple observations.

First, notice the right side of one shape lines up perfectly with the left side of the next shape, creating a continuous flow. This continuity suggests that perhaps these are not three separate shapes, but three shapes interconnected by a single concept.

Our focus is understanding the connection between these three curved shapes and why they all have an area of 1. First, let's label the bottom part of our combined shape by marking off units of 1. We'll start the left side of our blue curve at 1. Here are some of the numbers using our unit of 1 as our guide. Notice the far-right shape extends slightly past 20; we'll touch on the significance of that number in a moment.

| 1 | 2 | 3 | 4 | | 7 | | | 20 |

This setup provides some order and structure to our shapes, but it doesn't answer any of our "why" questions. It appears that the points where we change from one shape to another are arbitrary. We go from blue to red just before 3, from red to green just after 7, finally finishing just after 20. What is the significance to this observation?

$\frac{1}{1}$ $\frac{1}{2}$ $\frac{1}{3}$ $\frac{1}{4}$ $\frac{1}{7}$ $\frac{1}{20}$

| 1 | 2 | 3 | 4 | | 7 | | | 20 |

Our first discovery pertains to how we ascertain the height of the curve. It turns out, the height of the curve embodies a simple pattern: At 2, the height is $^1/_2$; at 3, the height is $^1/_3$; at 4 the height is $^1/_4$, continuing until we reach the height at 20, which is $^1/_{20}$. We defined the height at 1 as 1, which we can write as $^1/_1$. The key to understanding the pattern of this shape is to place it in the xy coordinate system where the bottom part is the x-axis. Then the top curve is a curve we generate by the equation $y = \frac{1}{x}$. Knowing this equation, we can determine this pattern is true for all input points x and not just integer values.

John Napier's story in the introduction illustrates the long, painstaking effort required to create logarithms. Thanks to him, we'll use this incredible tool to focus our story further. The previous graph summarizes all that Napier generated in his long research for the log table. As amazing as it may seem, every one of Napier's results is found in the image picture if we extend it far enough to the right.

Napier had a simple recipe: give me a number, and I will give you a natural log of that number. We have a calculator button for it, usually indicated by the *ln* button. For example, we can use our calculator to calculate the log of 2, or *ln*(2). The result is about 0.69314718. How this log function relates to our story is remarkable. The surprising relationship is the area under our curve between 1 and 2 is the same result, also about 0.69314718.

Area 0.693

1   2

Likewise, if we want to know the area between 1 and 3, we press 3 into our calculator and find it equals about 1.09861229. We can generalize this pattern for all cases. Therefore, we arrived at the first twist to our story: The area under a curve between 1 and some point x produces the same result as $ln(x)$.

Area 1.099

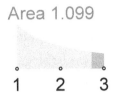

1   2   3

Our second twist arrives about a century later with Jacob Bernoulli, who delved into compound interest. He began with an optimistic scenario: doubling an investment in one year through compound interest. Then, he pondered the outcome if the one-year interval was divided into smaller periods. The first step is to consider the year in two equal parts. Instead of gaining 100 percent interest over a full year, he would earn 50 percent interest in each half, compounding midway through the year. So, if you invest $1 using this method, the amount after two compounding periods, or one year, is **1.5 × 1.5 × 1**, or 2.25. This was just the beginning.

If he divided the year into quarters, he would earn 25 percent each quarter. Then, after compounding after each quarter, the amount at the end of the year is 1 × 1.25 × 1.25 × 1.25 × 1.25 = 2.44. Bernoulli pursued this process to its extreme, envisioning the smallest possible time period that exceeds zero. In doing so, he determined that the final sum at the end of the year would fall between 2.71 and 2.72.

Area 1

1   2.718

How does this story of compound interest relate to John Napier and our picture? Notice that when you compute the logarithm of 2.71, you get a value slightly less than 1, while 2.72 yields a value slightly more than 1. If we were to use precise calculations instead of approximations, we would see that the exact value Bernoulli obtained through his compound interest computations corresponds perfectly to the value that results in a natural logarithm of 1. In other words, Bernoulli discovered this number that is approximately 2.71828 such that $ln(2.71828) = 1$, or the area under our curve between 1 and 2.71828 is 1.

This leads to our third twist with Leonard Euler about fifty years later. Bernoulli managed to piece together a remarkable discovery: the number 2.71828, accurate to five decimal places, holds a unique significance in mathematics. Euler deemed this number so special that he gave it a name—or rather, a symbol: e. This symbol is likely familiar to you as a function on your calculator, denoted as e^x. For instance, if you input the number 2 then e^x (which translates to e^2), the result will be approximately 7.389. If you input e^3, you'll get about 20.086. You may recognize these numbers as the numbers on our horizontal axis when we produced another area of 1 under our curve.

The picture below is a good summary of our discoveries. It is the same graph as before but to scale since the vertical axis is zoomed in to get a better picture of the area to the right. Notice the first number represents the input from the horizontal axis and the second number represents the value of the curve, or the output. Then each region, blue, red, and green, represents a region with area equal to 1.

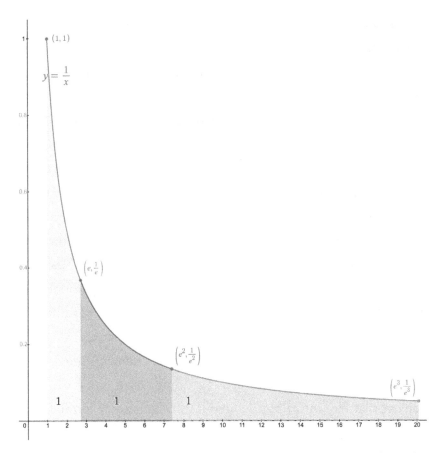

We've already determined that e is the specific value where the area under the curve equals 1. However, 7.389 (or more accurately, e^2) corresponds to an area of 2, and 20.086 (or e^3) aligns with an area of 3. This pattern extends across the entire curve. So, if we want to find the far right point where the area under the curve between 1 and that point equals 4, we calculate e^4, which results in roughly 54.6. This principle holds true not just for integer values, but for any given value.

What Euler uncovered, a fact overlooked for nearly 150 years, was that these two seemingly unrelated ideas were incredibly interconnected. Specifically, they are inverse operations, like addition and subtraction or multiplication and division.

Think of it this way: What is repeated addition? It is multiplication. What is repeated multiplication? It is exponentiation which means we use exponents. But what is the inverse of taking an exponent? In other words, how do we "undo" an exponent? We take the log. Perhaps you remember this connection. If we know $e^2 = 7.389$, then $\ln(7.389) = 2$, where ln corresponds to the base e.

Let's summarize our three main characters. Napier listened to the voice of pragmatic math to make it more efficient. Bernoulli listened to the voice of math that takes a business problem to its limit. Euler listened to the voice of math that brought all these ideas together and packaged it in an incredible way that makes sense of everything.

At the heart of this story is a way to transform multiplication into addition. This was Napier's goal, since multiplication was difficult, but addition was easy. What an amazing idea. Imagine thinking that there was a way, a method, to convert any two numbers into simple addition. That is the big idea of a logarithm, but it is lost when we convert that idea into a button on our calculator and a formula into a textbook.

How do we see this idea in our picture? Notice our picture starts at 1, which is the beginning point or identity in multiplication. In other words, $e^0 = 1$. Choose any two numbers greater than 1 as input. For simplicity, choose e^2 and e. Since that is our input, let's multiply them together. Because these are exponents, $e^2 e = e^3$. That is the result using multiplication.

Now we can see how we can solve for e^3 using addition. Our two inputs are e^2 and e. Find the result for e^2 on the graph, which is 2, and find the result for e on the graph, which is 1. Then, add the results: $2 + 1 = 3$. Then, to convert this to a multiplication result, find the result 3 on the horizontal axis and identify what input produces an output of 3. That result is e^3, which matches the result of multiplication.

This example may not have appeared to be easier because we chose the special input numbers with base e. But, let's repeat the process not knowing this shortcut and simply multiply the numeric values. The numeric value for e is about 2.718 and the numeric value for e^2 is about $2.718^2 = 7.389$. So, multiplying 2.718 by 7.389 without a calculator is a tedious and time-consuming process. But, we know $ln(2.718) = 1$ and $ln(7.389) = 2$. Since $1 + 2 = 3$, to get the result, we just need to look in the log tables to find $ln(20.086) = 3$ so that indicates that $2.718 \times 2.718^2 = 2.718 \times 7.389^{\square} = 20.086$.

Here is this same idea in pictures.

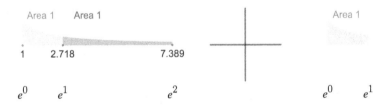

Results in

We transformed multiplication into addition. Of course, this opens up an entire course of options. This works for any combination we dream up, such as multiplying 4.481689 and 12.1824940.

Since $ln(4.481689) = 1.5$ and $ln(12.1824940) = 2.5$ and $1.5 + 2.5 = 4$, to get the result, we just need to look in the log tables to find $ln(54.598150) = 4$ so that indicates that $4.481689 \times 12.1824940 = 54.598150$ (within rounding).

Without calculating, what is the square root of 7.389? What is the square of 7.389? Without calculating, would you agree that $\ln(3) + \ln(4) = \ln(12)$?

And questions for the adventurous crowd: clearly, the area under our curve is diminishing quickly as we travel left to right. Notice how much further to the right we must travel to go from an area of 2 to 3 to 4, etc. Is there a limit as to how much area we can accumulate? Can we accumulate an area of 10 or 100 or 1,000 under the curve by travelling farther right along the x-axis? Are you guessing or are you certain?

Also, I hope next time you click the ln button on your calculator, you think about the beautiful relationships of area under a curve that are happening behind the scenes.

Chapter 1: From Here to Eternity

Let's take the square from the previous story and arrange it as depicted in the top left of Figure A.1. Now, consider adding more of the same squares so we produce the configuration in the top right of Figure A.1. Notice this configuration is $2 \times 2 = 2^2 = 4$ squares. Let's consider how many squares we added to the first configuration in order to create the second configuration. We started with one square and we finished with four squares, so the number of squares added is three.

OK, let's produce the next configuration which is the bottom left that contains nine squares ($3 \times 3 = 3^2 = 9$). Now, repeat our question: how many squares did we add from configuration two to produce configuration three? We started with four squares and finished with nine so that required five more squares.

Let's repeat once more to produce the fourth configuration at the bottom right as sixteen squares ($4 \times 4 = 4^2 = 16$). The number of squares required to produce this configuration from the previous is $16 - 9 = 7$.

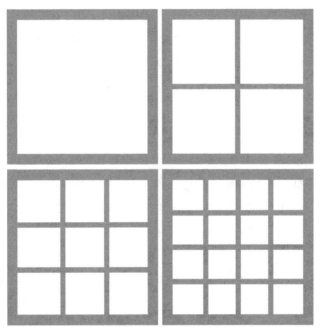

Figure A.1: *Increasing the size of a shape by the factor of a single positive integer squares the overall size.*

Do you notice any pattern? In order, the number of squares we added to produce the next configuration is three, five, and seven. If we assume we started with no squares, then we add one square to produce the one-square configuration. This produces our pattern of adding squares to produce the next perfect square as one, three, five, and seven. Notice each of these are odd numbers. Not just any odd number but consecutive odd numbers.

Can you identify how this pattern works?

Let's first identify why we always need to add an odd number of squares to produce the next configuration. The secret is the symmetry at play with the rows and columns of a square. Consider the process of going from a 3×3 configuration to a 4×4 configuration. Let's do this in three steps.

First, create a fourth column to the right by adding three squares, which creates a 3×4 configuration.

Second, create a fourth row at the bottom by adding three squares.

The third step is to add the one diagonal square at the bottom right.

Notice the pattern is that whatever number of squares we add in the first step we repeat in the second step. In this example, we added three in step one and step two which is a total of $3 \times 2 = 6$ squares added in the first two steps. Because the first step always adds the same number as the second step, we have a symmetry that results in always multiplying that number by 2. Any positive integer multiplied by 2 is an even number. Then, in the final step, we always add one square for the final diagonal. Adding 1 to an even number always produces an odd number.

I will leave it as an exercise for you to identify why the pattern is the next odd number.

The fascinating take-away here is that we created a pattern of adding the next odd number to create the next larger perfect square configuration, which relates odd numbers to squares. At first, this seems to be an odd connection, but now you know the secret!

Chapter 2: In the Realm of Nothingness

In the previous story, we created larger squares by increasing the side lengths by one unit. Let's consider a square that has a side length of 2 and then increase the side length by a very small amount. We could choose some small length, such as 0.001 but let's use the notation dx to represent this small increase. So, think of dx as a small number, such as 0.001 or 0.0001. Then, the question we want to consider is how does the area of a square change based on increasing the side length from 2 to $2 + dx$?

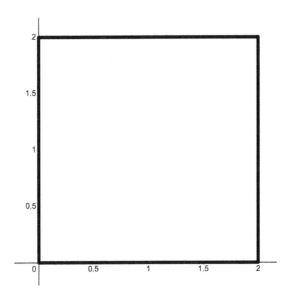

First, the picture above is a square with a side length of 2. We can identify how the area changes by increasing the side length to $2 + dx$ by modifying this square based on the next picture.

Because dx represents an unknown small number, we treat it like we would the variable x in algebra. The length dx in the picture is larger than we would imagine but choosing this larger size allows us to visualize the length.

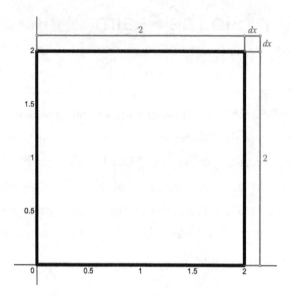

Now, use this picture to visualize the increase in area that results by increasing the side length by dx. Notice at the top we added a rectangle with dimensions 2 and dx so the total area added to this rectangle is $2dx$. We have another rectangle on the right that is also 2 by dx with area $2dx$ so we added two rectangles with area $2dx$, which means the total area of the two rectangles is $4dx$.

But we also added a small square with length dx so its area is $dx \times dx = (dx)^2$. Remember dx represents a very small number greater than 0. This is where things get interesting and we must use our imagination.

Imagine dx getting smaller and smaller and approaching 0 but never quite reaching 0. The creative rule calculus developed in this process is to assume we use all factors of dx but we discard any factor that is a higher power of dx. In this case, ignore the grayed-in $(dx)^2$ term area. Vanish. Gone.

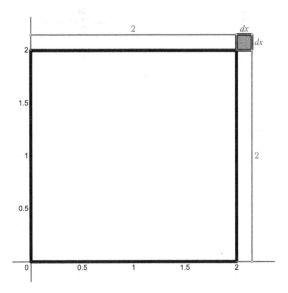

This may seem like an illegal move because it appears an arbitrary rule. This is the challenge Newton and Leibniz wrestled with. Math has developed stronger logic regarding this rule that is beyond the scope of this book. The hope here is we can at least consider this as a curious rule. When we apply this rule, then the increase in area for a square of length 2 becomes $4dx$.

We can generalize this result and consider a square that has length x. Then, if we increase the side length to $x + dx$, we increase the area from x^2 to $x^2 + 2x\ dx$. This leads to the conclusion that for a square with side length x, if we increase the length by dx, we increase the area by $2x\ dx$.

It is easy to consider a square as a static object. However, this highlights that a beautiful part of math is changing our perspective from a simple static object, such as a square, and imagining how the properties of the square change as the length of the side changes. It's like mathematicians are always asking the question "What if?"

Chapter 3: Beneath the Surface

In our introduction story, we explored different representations of shapes that have an area of 1. The simplest is a square with a side length of 1. We could place the square on the xy plane with the bottom left corner at the origin which means the top right corner is at the point $(1,1)$. Then we identified the graph $y = \dfrac{1}{x}$ and considered the area below this graph and above the x-axis between $x = 1$ and $x = e$. This area is also 1. Recall we have a calculator button that reproduces this result when we enter the value for x at the far right point, which is e, then press the ln button to produce the result 1. The picture below illustrates two different regions that have an area of 1.

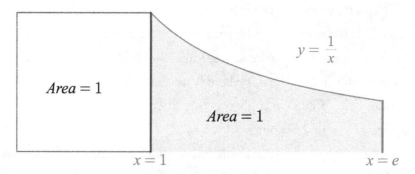

In chapter 3, we considered the negative numbers so let's identify an area of 1 that is connected to the negative numbers. Recall the curve $y = e^x$. Now, consider the area under this curve and above the x-axis for the negative x values.

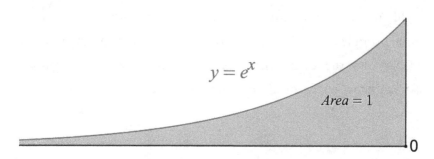

The graph above is the graph $y = e^x$ for all negative values for x. It may be difficult to consider the graph for all negative numbers since the negative numbers extend to infinity going right to left on the x-axis, which means our shape does not have a beginning on the left side. In the previous chapter story, we considered the case where dx approached 0 but never actually arrived at 0. Here, we consider the case where x approaches negative infinity but never actually arrives there. So, again we rely on our imagination to envision this limit. In this limiting assumption, the area below the curve from negative infinity to 0 approaches the number 1. This means we can let the area get as close to 1 as we desire but it will never reach or exceed 1 regardless of which negative number we start with.

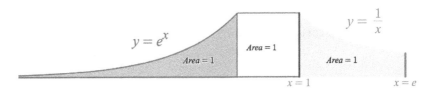

In the graph above, the left region represents an area of 1 when we consider the limit as x approaches negative infinity. Now when we view our entire collection, the total area is 3. The area to the left of $x = 0$ is 1. The squared area between 0 and 1 is also 1. Finally, we get our third area of 1 between $x = 1$ and $x = e$.

Notice in this creative collection, we use e as an input in the area on the right as it is the rightmost value for x. We also use e in the left region but it is part of the output as the base value for our curve. e often appears when we consider the area under a curve which is why we will consider a different perspective of e in our next story.

Chapter 4: The Fractional Realm

In the previous story, we learned how e uses its exponential powers by receiving the entire spectrum of the negative numbers and by the surprise process of calculating the area under the curve of this exponential process returns the number representing unity, which is 1. But we've also learned how starting at 1 and dividing the positive numbers into intervals of a factor of e also calculates the unity of 1 below the curve $y = \dfrac{1}{x}$. Not surprisingly, e is an irrational number which means we cannot express it in terms of a finite number of rational numbers. However, as another surprise twist in the novel of e, we can express it as a sum of rational numbers but it is an infinite sum.

Here are the first six terms in the infinite sum for e: $\dfrac{1}{0!} + \dfrac{1}{1!} + \dfrac{1}{2!} + \dfrac{1}{3!} + \dfrac{1}{4!} + \dfrac{1}{5!}$. The exclamation point indicates the factorial function, which is simply repeated multiplication. $0! = 1$ by definition. $1! = 1$, $2! = 2 \times 1 = 2$ and $5! = 5 \times 4 \times 3 \times 2 \times 1 = 120$. Think of this as an infinite walk where each step lands on the next dot. Thus, the distance we travel in each step is the distance between each dot. So, start at 0 and step right one unit to travel from 0 to 1 since $\dfrac{1}{0!} = 1$. Likewise, since the distance of the next step is $\dfrac{1}{1!} = 1$, in the second step we go from 1 to 2. Once at 2, the next distance is $\dfrac{1}{2!} = 0.5$, so in the third step, we travel from 2 to 2.5. Here is a picture of the first three steps, where we travel a total distance of 2.5.

DISTANCE BETWEEN POINTS

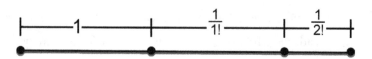

Now let's picture the next three steps which are the first six terms. Because the last two terms are so small, we had to zoom in and picture the final two terms in a separate zoomed-in picture.

DISTANCE BETWEEN POINTS

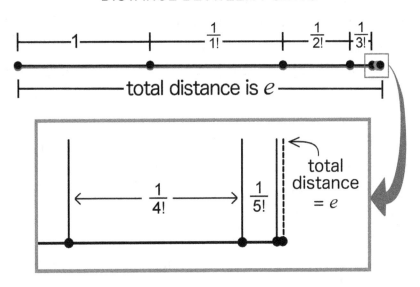

Even in the zoomed view, the distance traveled after the sixth step is nearly at the destination e. Isn't it remarkable that we can express e by this simple, yet beautiful pattern? In order to reach the full status of an irrational number, e requires every single positive integer as input, which is an infinite set.

The entire pattern follows the pattern we started where the numerator is always 1. The nth term is $\frac{1}{(n-1)!}$. Here is how we would write this sum: $\frac{1}{0!} + \frac{1}{1!} + \frac{1}{2!} + \frac{1}{3!} + \cdots$. Technically, we would state that as the number of terms approaches infinity, the result of the sum approaches e.

You may wonder how quickly this sum approaches e. If there are an infinite number of terms, then the fifty-third term is barely beginning.

The fifty-third term is $\frac{1}{52!}$. In order to understand how small this term is, we can consider how large $52!$ is. One way to think of $52!$

is that it represents the total number of ways to uniquely shuffle a fifty-two-card deck of playing cards. This seems like a large number, but how large?

As strange as it may seem, this number is greater than the number of stars in the universe. That gives you an idea as to how small the number $\frac{1}{52!}$ is and it is just the beginning of the infinite number of terms. Imagine how precise our estimate would be of e if we only included the first fifty-three terms? The decimal precision is greater than the number of stars in the universe. But we are only getting started.

The value of the factorial notation is that it is compact. But this compact notation is misleading because it hides the insane size the results are for relatively small integers, such as 52.

Hopefully this highlights that the irrational number e is fundamentally different than rational numbers. It is not just more of the same but it is a number defined by patterns or properties.

Because each term in this series is a rational term and a finite sum of rational numbers is rational, then the sum of the first fifty-three terms would be a rational approximation for the true value of e. What a wonderful gift we've been given to be able to understand this pattern even if the size of it is beyond measure.

Chapter 5: The Irrational Unknown

In the previous story, we viewed the irrational number e in a rational perspective. This section is about irrational numbers and since π is the most celebrated irrational number, let's explore fresh ways to think of π.

You may recall the formula for the circumference of a circle as $2\pi r$ where r is the length of the radius. Another way to say this is if the circle has radius of 1, then the distance around the circle is 2π. You likely have solved a problem in math before where you needed to calculate the circumference. Let's not just calculate but rather compare these two distances. Let's compare these two distances on an apples-to-apples basis by flattening the distance around the circle into a straight segment as we do in the figure below.

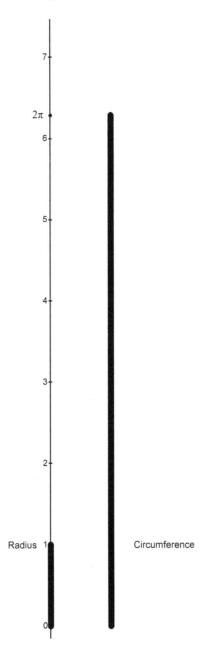

Next, compare this circle to another circle where we double the radius to 2. When we double the radius, what impact will that have on the circumference? Now, the circumference is 4π. Here is a picture of the two comparisons.

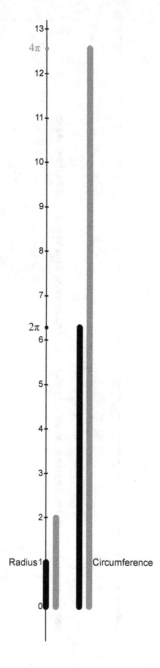

What do you notice? It appears that as we doubled the radius, we also doubled the circumference. In fact, this is exactly what occurs. Have you ever thought about how the change in the radius of a circle impacts the change in the distance around the circle? Now you can visually see that change. This suggests that if we triple the radius, we triple the circumference.

Now, let's change perspectives by rotating the radius length to be horizontal. We placed each circle on a separate *xy*-coordinate system where the radius starts at the origin which is labeled as point A. Then it goes right to the point R_1 for the radius of 1 and R_2 for the radius of 2. Then the circumference for the radius of 1 is the vertical line that starts at R_1 and goes up to point C_1. We repeated this method for the radius 2 circle. Notice the two diagrams appear to be the same size but the diagram on the right is compressed by a factor of ½.

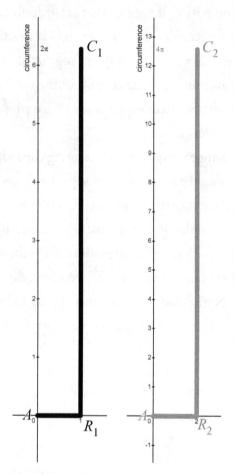

That means that if we change comparison from the black graph of radius 1 to the green graph of radius 2, the comparison appears identical if we zoom out and ignore scale. Our next move may appear odd but let's draw a segment to connect the points R_1 and C_1 and the points R_2 and C_2.

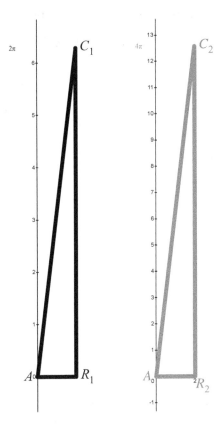

Now we've created two triangles, AR_1C_1 and AR_2C_2. Again, the triangles appear the same but that is because we compressed the graph on the right. Specifically, these two triangles are similar triangles in that they have the same shape but different sizes.

Next, let's consider the equation for the line that connects A to either C_1 or C_2. You may recall from algebra the formula for a line is $y = mx + b$.

The slope is rise over run, which is $\frac{2\pi}{1} = 2\pi$ for the radius 1 graph and $\frac{4\pi}{2} = 2\pi$ for the radius 2 graph. So, both have the same slope and both have the same y-intercept of 0. Thus, the formula for both is $y = 2\pi x$. Since y represents the circumference and x represents the radius, we can rewrite this equation as $C = 2\pi r$. This is not only true for these two circles but any circle. That means we can draw any right triangle that uses this line as the hypotenuse, and this will create a triangle that models the relationship between the radius and circumference.

One interesting take-away here is the circumference is a linear function of the radius. We recognize linear functions in the form $y = mx + b$. Since the y-intercept is 0, the formula simplifies to $y = mx$. The slope is rise over run where rise is the circumference and run is the radius.

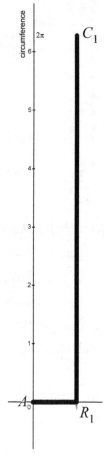

Notice the shift in perspective. Now we can compare the radius and circumference of a circle by using triangles rather than segments. If we pause a moment, we can appreciate the significance that we are using an object with straight lines, a triangle, to model the relationship of a curved object, a circle. It's like how thirty years ago, we thought of a phone as a device used to verbally communicate with one other person. Now, with the smartphone, we have a much wider perspective of what a phone is. We can think of this type of triangle as a "smart triangle" where it also models the relationships of a circle.

But, there is more magic to these "smart" triangles than just modeling the linear relationship between the radius and circumference. There is a surprising relationship hiding in the second dimension.

Notice our linear relationship is a distance relationship. But we can also increase the dimension from a one-dimensional distance relationship to a two-dimensional relationship that involves area.

Let's choose a circle that has a generic radius r and model this circle with our same "smart" triangle. The triangle has base r and the height is the circumference of a circle with radius r, or $2\pi r$. Since we've created a right triangle, we can easily calculate the area for this triangle as $\frac{1}{2} base \times height = \frac{1}{2}r \times 2\pi r = \pi r^2$.

Recall πr^2 matches the area within the circle, which is the two-dimensional object called a disk, which is the outer circle and the area within the circle. Our surprise is how the area for a triangle models the area for a circle. Here is a picture of the circles and triangles together.

Let's summarize what we've learned by first analyzing the black figures. The radius of the circle matches the length of the base to the triangle. The circumference length matches the height of the triangle (dotted lines). Then the area of the triangle also matches the area of the circle. This not only is true for the black figures but the green figures whose circle has a radius twice the length of the black circle.

Isn't this a more interesting way to view math rather than memorizing the formulas for the circumference and area of a circle?

Chapter 6: An Imaginary Universe

In our normal xy-coordinate system, we can take an equation that involves x and y and think about x as our input and y as our output. In our previous story, the input is the radius r and the output is the circumference C. Here is a graph of this equation.

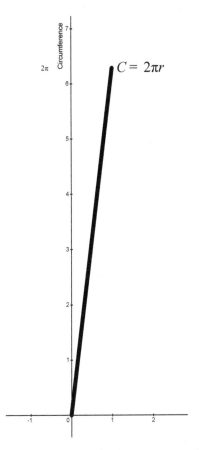

The natural perspective is both the x-axis and y-axis are represented by a straight line. We've learned in this book to expand our thinking to different dimensions, and we will continue that mental stretching here. Previously in our story from section 3, we considered the equation $y = e^x$ and fed all the negative real numbers into this equation which produced this curve.

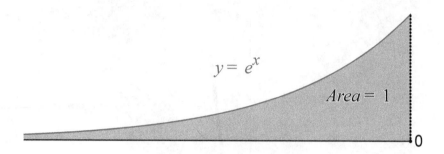

How does this function change if we multiply x by the imaginary number i in the exponent? In other words, how does the graph change if our equation becomes $y = e^{xi}$?

Unless you are used to working with these types of equations, this likely seems to be an odd question at best. Remember the definition is $i^2 = -1$ or $i = \sqrt{-1}$. We don't have a button on our calculator for a square root of a negative number. Not only that but placing i in the exponent makes it even more difficult to interpret. But it has a very surprising, and beautiful interpretation.

In the normal real-number xy-coordinate system, we graph a function by plotting input/output points from the function. For example, given the function $y = e^x$, an $x = 2$ input produces an output $y = e^2$. Then, we plot the pair $(2, e^2)$ in the xy plane. But, the imaginary number i introduces rotation into the process which changes the usual straight line process we are used to.

The surprise is when we input some real number x into the function $y = e^{xi}$ and plot the result on a complex plane, the output is always a point on a circle with a radius of 1 where the center of the circle is at the origin. Let's plot a few points to illustrate how this process works.

We can think of $x = 0$ as the starting point. The output for this input is the point $(1, 0i)$. From this starting point, we can easily identify other points. The process is to assume the input x represents the distance travelled along the unit circle in a counterclockwise direction.

To get our bearings, assume we input $x = 2\pi$ into our function. In order to determine the location of the output point, start at the point $(1,0i)$ and travel a distance 2π counterclockwise around the circle. Since the circumference of a circle with radius 1 is 2π, then the distance 2π represents a complete journey around the circle so the output is the same point $(1,0i)$.

Once we have that identified, it is easy to plot other points. For example, if we input $x = \pi$, we traveled halfway counterclockwise around the circle and the location is on the other side of the circle with point $(-1,0i)$.

Here are the results for these two values (solid black dots) as well as the integer values 1–6 (the black dots that are not solid).

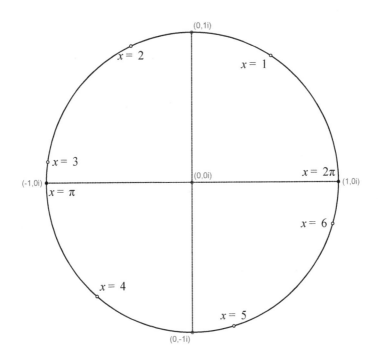

If it helps, you can think of this process similar to a 400-meter oval track field. There is a starting point on a track that would represent $x = 0$. Then, if we ran 100 meters on the track, we would be ¼ of the way around the oval. If we ran 400 meters, our location would be the same location as where we started. Here we have a circle rather than an oval and the distance around is 2π rather than 400 meters but the process is the same.

One nice feature of the complex plane is we can write each point as a complex number. A complex number is a number that can be expressed in the form $a + bi$, where a and b are real numbers. For example, the point $(3, 4i)$ in the coordinate system is the complex number $3 + 4i$. So, in our previous example, the input $x = \pi$ produces the output location point $(-1, 0i)$. Then, to convert the output point to a number, $a = -1$ and $b = 0$, which produces the number as $-1 + 0i = -1$.

That means we can write our result as $y = e^{xi} = e^{\pi i} = -1$. In other words, we identified $e^{\pi i} = -1$. This equation is often hailed as the most beautiful equation in mathematics because it links together three of the most important constants in mathematics: e (the base of the natural logarithm), π (the ratio of a circle's circumference to its diameter), and i (the imaginary unit). These numbers originate from completely different sources. But amazingly, we can combine them into one expression and the result simplifies to -1.

At times, math can appear to be like magic. One major difference between math and magic is when we understand the secret behind a magic trick, there can be a letdown. But, when we understand why a seemingly magical result appears in math, it increases our appreciation because it only opens the door to deeper mysteries not yet solved.

REFERENCES

Archimedes, Reviel Netz, and Ascalonita Eutocius. *The Works of Archimedes*. Cambridge: Cambridge University Press, 2004.

Arndt, Gary. "The history of negative numbers." Everything Everywhere. March 24, 2023. https://everything-everywhere.com/the-history-of-negative-numbers/.

Beasley, John D. *The Mathematics of Games*. Mineola, NY: Courier Corporation, 2013.

Bell, Eric Temple. *Development of Mathematics*. New-York: McGraw-Hill, 1945.

Bell, Eric Temple. *The Development of Mathematics ... Second Edition*. New York: McGraw-Hill Book Company, 1945.

Best Infographics. "The drake equation." January 25, 2021. https://www.best-infographics.com/the-drake-equation-infographic/.

Bostock, Mike. "Ulam spiral." Observable. April 26, 2023. https://observablehq.com/@mbostock/ulam-spiral.

Boyer, Carl B., and Uta C. Merzbach. *A History of Mathematics*. Hoboken, New Jersey: John Wiley & Sons, 2011.

Britannica. "Foundations of mathematics - reexamination, infinity, axioms | Britannica." www.britannica.com. Accessed August 29, 2023. https://www.britannica.com/science/foundations-of-mathematics/The-reexamination-of-infinity.

Calculator.net. "Interest calculator." Calculator.net. 2019. https://www.calculator.net/interest-calculator.html.

Day, Dustin. *Apple under Microscope*. November 8, 2010. *Flickr*. https://www.flickr.com/photos/pastedtoast/5156644766/.

Dennis, Geoffrey. "Judaism and Numbers." My Jewish Learning. October 16, 2008. https://www.myjewishlearning.com/article/judaism-numbers/.

Descartes, René. *The Geometry of René Descartes*. Chicago, IL: Open Court, 1925.

Euclid. *Euclid's Elements of Geometry*. Translated by Richard Fitzpatrick, 2008. https://farside.ph.utexas.edu/Books/Euclid/Elements.pdf. Accessed April 9, 2024.

Folkerts, Menso. "Mathematics - mathematics in ancient Egypt." Encyclopædia Britannica. 2019. https://www.britannica.com/science/mathematics/Mathematics-in-ancient-Egypt.

Hobson, Ernest William. *John Napier and the Invention of Logarithms, 1614*. Cambridge, England: Cambridge University Press, 1914.

Hooper, Dan. *At the Edge of Time*. Princeton, NJ: Princeton University Press, 2021.

Horenberger, Beau. "Egyptians and the Rhind mathematical papyrus: historical context (1)." The Horenberger Zone. May 26, 2022. https://horenbergerb.github.io/2022/05/26/egyptianmath1.html.

Ifrah, Georges. *The Universal History of Numbers*. London: Harvill Press, 1998.

Ifrah, Georges. *The Universal History of Numbers.* Hoboken, NJ: Wiley, 2000.

Kaplan, Robert. *The Nothing That Is: A Natural History of Zero.* Oxford; New York: Oxford University Press, 2000.

Kibe, Tanay, Sukrut Mondkar, Ayan Mukhopadhyay, and Hareram Swain. "Black Hole Complementarity from Microstate Models: A Study of Information Replication and the Encoding in the Black Hole Interior." *Journal of High Energy Physics* 96 (July 10, 2023). https://doi.org/10.1007/jhep10096.

Kiger, Patrick. "What is Planck's constant, and why does the universe depend on it?" HowStuffWorks. December 10, 2019. https://science.howstuffworks.com/dictionary/physics-terms/plancks-constant.htm.

Lewin, Christopher. "The Emergence of Compound Interest." *British Actuarial Journal* 24 (2019): E34. doi:10.1017/S1357321719000254.

Lin, Pi-Jen, and Wen-Huan Tsai. "Enhancing Students' Mathematical Conjecturing and Justification in Third-Grade Classrooms: The Sum of Even/Odd Numbers." *Journal of Mathematics Education* 9, no. 1 (June 2016): 1–15.

Lockhart, Paul. *Measurement.* Cambridge, MA: Harvard University Press, 2012.

MasterClass. "String theory explained: a basic guide to string theory." MasterClass, June 7, 2021. https://www.masterclass.com/articles/string-theory-explained.

Mastin, Luke. "Egyptian mathematics - numbers & numerals." The Story of Mathematics - A History of Mathematical Thought from Ancient Times to the Modern Day, 2020. https://www.storyofmathematics.com/egyptian.html/.

Mathematics Magazine. "The Sumerian mathematical system." mathematics-magazine.com. Accessed August 21, 2023. http://mathematicsmagazine.com/Articles/TheSumerianMathematicalSystem.php.

Mumford, David. "What's so Baffling about Negative Numbers? – a Cross-Cultural Comparison." In *Studies in the History of Indian Mathematics*, edited by C. S. Seshadri, 113–143. Gurgaon: Hindustan Book Agency, 2010.

Nahin, Paul J. *An Imaginary Tale*. Princeton: Princeton University Press, 2010.

Netz, Reviel. *A New History of Greek Mathematics*. Cambridge; New York: Cambridge University Press, 2022.

Ore, Øystein. *Number Theory and Its History*. New York: McGraw-Hill Book Co., 1948.

Osterloff, Emily. "How an asteroid ended the age of the dinosaurs." Nhm.ac.uk. November 18, 2020. https://www.nhm.ac.uk/discover/how-an-asteroid-caused-extinction-of-dinosaurs.html.

Ouellette, Jennifer. "The fuzzball fix for a black hole paradox." Quanta Magazine. June 23, 2015. https://www.quantamagazine.org/how-fuzzballs-solve-the-black-hole-firewall-paradox-20150623/.

Our World in Data. *Population, 10,000 BCE to 2021*. 2021. *Our World in Data*. https://ourworldindata.org/grapher/population.

Pisano, Leonardo, and Laurence E. Sigler. *Fibonacci's Liber Abaci: A Translation into Modern English of Leonardo Pisano's Book of Calculation*. New York: Springer, 2003.

Popova, Maria. "The invention of zero: how ancient Mesopotamia created the mathematical concept of nought and ancient India gave it symbolic form." The Marginalian. February 2, 2017. https://www.themarginalian.org/2017/02/02/zero-robert-kaplan/.

Quote Investigator. "Tortoises all the way down – Quote Investigator®." Quote Investigator. August 22, 2021. https://quoteinvestigator. com/2021/08/22/turtles-down/.

Redd, Nola Taylor. "How was Earth formed?" Space.com. Space.com. November 1, 2016. https://www.space.com/19175-how-was-earth-formed.html.

Resourceaholic. "New GCSE: tangents and areas." Resourceaholic, n.d. https://www.resourceaholic.com/2015/08/graphs.html.

Rogers, Leo. "The history of negative numbers." NRICH. University of Cambridge. 2009. https://nrich.maths.org/5961.

Rooney, Anne. *The Story of Mathematics*. London, UK: Arcturus, 2014.

Saito, Ken. "Doubling the Cube: A New Interpretation of Its Significance for Early Greek Geometry." *Historia Mathematica* 22, no. 2 (May 1995): 119–137. https://doi.org/10.1006/hmat.1995.1013.

Scott, Joseph Frederick. "John Napier | Scottish Mathematician." Encyclopædia Britannica. March 31, 2019. https://www.britannica.com/biography/John-Napier.

sites.google.com. "The renaissance - mathematics." n.d. https://sites.google.com/a/stu.sandi.net/the-renaissance/science-and-inventions/mathematics.

Stapel, Elizabeth. "Introduction to negative numbers | Purplemath." Purplemath. 2019. https://www.purplemath.com/modules/negative.htm.

Stein, Howard. "Eudoxos and Dedekind: On the Ancient Greek Theory of Ratios and Its Relation to Modern Mathematics." *Synthese* 84 (1990): 163–211.

Stillwell, John. *The Story of Proof.* Princeton, NJ: Princeton University Press, 2022.

Sutori. "A brief history of negative numbers." www.sutori.com. n.d. https://www.sutori.com/en/story/a-brief-history-of-negative-numbers--p5qb-nmwDGKQ3TbTYRqJToHHw.

Tattersall, James J. *Elementary Number Theory in Nine Chapters*. New York: Cambridge University Press, 1999.

Therieau, Lillie. "Brahmagupta: the man who defined zero." www.ele-phantlearning.com, n.d. https://www.elephantlearning.com/post/brahmagupta-the-man-who-defined-zero.

Ye, Xiaojing. "Pi day: how did they first calculate pi?" Christian Science Monitor. March 14, 2016. https://www.csmonitor.com/Science/2016/0314/Pi-Day-How-did-they-first-calculate-pi.

ACKNOWLEDGMENTS

As I penned the pages of this book, I was compelled to reflect on the myriad of individuals who have shaped my journey to the present. This reflection brought into sharp focus the gratitude I owe to those who have been integral to my story. While it's impossible to acknowledge each person by name, the narrative begins with my family, whose love endures despite an intimate acquaintance with my flaws.

My path has been significantly influenced by an array of teachers, coaches, pastors, and mentors. Often, it's the smallest gestures—an unexpected kindness, the bravery to confront life's adversities, or thankfulness in the face of hardship—that have left the deepest imprint on me. I recall the early days with the staff and clients at SALT Solutions, whose perseverance through challenging times was nothing short of inspirational. I am equally thankful for the expanded team at Coaching Actuaries and our customers, who have made it possible for me to immerse myself in mathematics and deem it my profession.

I cherish the moments spent with those in our mathematics book clubs, Roots and Wave, and I'm grateful for the IntersectingUs team, driven by a mission to effect meaningful change in people's lives.

My appreciation extends to those who have nurtured and preserved the legacy of mathematics, bringing us to our current understanding. I remain optimistic about the future, anticipating the thrill of uncovering new and thrilling discoveries in mathematics.

As I think about the mega story of mathematics, I'm reminded of our recent solar eclipse. It was an awe-inspiring celestial event that illustrated the beauty and precision of the universe. However, the solar eclipse also captivated us with the illusion of the sun and moon appearing equal in size, serving as a poignant reminder of how mathematics reveals truths beyond our visual perceptions.

And yet our perceptions are part of our reality. Understanding occurs when we combine our perceptions with the truth of math. For this, I am grateful to the Voice of math that unveils all truth.

Logic can be patient because it is eternal.
—OLIVER HEAVISIDE

ABOUT THE AUTHORS

DAVE KESTER

Dave's career is a testament to the art of transforming obstacles into stepping stones. He holds a firm belief in the potential of change to open new doors, attributing many of his pivotal decisions to unforeseen shifts. This philosophy permeates every facet of his life, enriching both his personal journey and his teachings.

Much of Dave's professional life has been dedicated to the intricate world of actuarial exams where he has been a beacon for numerous aspiring actuaries as the founder of Coaching Actuaries. Although he recently stepped down as President, he continues to shape the company's future as Chairman of the Board, focusing on content development and customer success. Dave is also the driving force behind IntersectingUs.com, a platform he established to inspire a deeper appreciation of the interplay between mathematics and life's broader perspectives.

Dave's dedication extends beyond the professional sphere; he is committed to building connections and supporting individuals and teams in their pursuit of goals within mathematics, actuarial science, family, and faith.

MIKAELA ASHCROFT

Mikaela has a master's degree in physics and has worked as a ghost-writer and author for over thirteen years. Physical and mental exploration of the unknown has always been among the most important goals for Mikaela, and the collaborative effort with Dave has proved among the most enlightening yet. Enjoy this dip into the unknown, and perhaps you'll find yourself along the journey.